The Blind Spot

T0357198

The Blind Spot

Why Science Cannot Ignore Human Experience

Adam Frank, Marcelo Gleiser, and Evan Thompson

The MIT Press
Cambridge, Massachusetts
London, England

First MIT Press paperback edition, 2025

The MIT Press would like to thank the anonymous peer reviewers who provided comments on drafts of this book. The generous work of academic experts is essential for establishing the authority and quality of our publications. We acknowledge with gratitude the contributions of these otherwise uncredited readers.

This book was set in ITC Stone Serif Std and ITC Stone Sans Std by New Best-set Typesetters Ltd. Printed and bound in the United States of America.

Library of Congress Cataloging-in-Publication Data

Names: Frank, Adam, 1962- author. | Gleiser, Marcelo, author. | Thompson, Evan, author.
Title: The blind spot : why science cannot ignore human experience / Adam Frank, Marcelo Gleiser, and Evan Thompson.
Description: Cambridge, Massachusetts : The MIT Press, [2024] | Includes bibliographical references and index.
Identifiers: LCCN 2023012137 (print) | LCCN 2023012138 (ebook) | ISBN 9780262048804 (hardcover) | ISBN 9780262377751 (epub) | ISBN 9780262377744 (pdf), 9780262553032 (pb.)
Subjects: LCSH: Science—Philosophy. | Metaphysics. | Cosmology—Philosophy. | Philosophy of mind.
Classification: LCC Q175 .F768 2024 (print) | LCC Q175 (ebook) | DDC 501—dc23/eng/20231031
LC record available at https://lccn.loc.gov/2023012137
LC ebook record available at https://lccn.loc.gov/2023012138

10 9 8 7 6 5 4

Contents

An Introduction to the Blind Spot

The Fire This Time

We write this book with a sense of urgency because we believe our collective future and human project of civilization are at stake. The success of modern science at gaining knowledge of nature and control over it has been spectacular. Think, for instance, of the recent development of mRNA vaccines, an entirely new kind of vaccine with huge implications for treating many diseases. At the same time, science denial is on the rise and civil society is splintered. Most threatening of all, our scientific civilization confronts an inescapable calamity of its own creation, the planetary climate crisis. If we cannot find a new path forward, our globe-spanning civilization and all who depend on it may be unable to cope with immense challenges.

We believe we need nothing less than a new kind of scientific worldview. Since the dawn of the Enlightenment, we have increasingly looked to science to tell us who we are, where we come from, and where we're going. In the seventeenth century, a new worldview that was intimately connected to the rise of modern science but not identical to it began to spread. By the nineteenth century, it had turned into a juggernaut, transforming culture and its material basis faster than at any other time in human history. According to that worldview, nature is nothing but shifting spatiotemporal arrangements of fundamental physical entities. In this perspective, the mind is either a derivative physical assemblage or something radically different from nature altogether. Most important, science gives us a literally true account of objective physical reality or at least of the totality of observable physical facts. This worldview of nature, mind, and science eventually came to underpin our political systems, our economic structures, and our social organization. But it is precisely that philosophical perspective, including its presence within scientific theories themselves, that is now in crisis, as evidenced by its inability to account for the mind, meaning, and consciousness that are the very source of science itself. In the wake of an extended series of puzzles and paradoxes, which we chart in the chapters in this book, about time and the cosmos, matter and the observer, life and sentience, mind and meaning, and the nature of consciousness, we have been left unsure of how to make sense of ourselves and our place in the world. Worse still, the crisis in our understanding occurs at a crucial moment in history when we face multiple existential challenges, such as climate change, habitat destruction, new global pandemics, digital surveillance, and the growing prevalence of artificial intelligence, all of them driven by the success of science and technology. The COVID-19 pandemic brought the urgency of these concerns to the forefront as we have experienced the fragility of our species in a natural world that we cannot and should not relate to simply as a material resource to control.

Our scientific worldview has gotten stuck in an impossible contradiction, making our present crisis fundamentally a crisis of meaning. On the one hand, science appears to make human life seem ultimately insignificant. The grand narratives of cosmology and evolution present us as a tiny contingent accident in a vast indifferent universe. On the other hand, science repeatedly shows us that our human situation is inescapable when we search for objective truth because we cannot step outside our human form and attain a God's-eye view of reality. Cosmology tells us that we can know the universe and its origin only from our inside position, not from the outside. We live within a causal bubble of information—the distance light traveled since the big bang—and we cannot know what lies outside. Quantum physics suggests that the nature of subatomic matter cannot be separated from our methods of questioning and investigating it. In biology, the origin and nature of life and sentience remain a mystery despite marvelous advances in genetics, molecular evolution, and developmental biology. Ultimately, we cannot forgo relying on our own experience of being alive when we seek to comprehend the phenomenon of life. Cognitive neuroscience drives the point home by indicating that we cannot fully fathom consciousness without experiencing it from within. Each of these fields ultimately runs aground on its own paradoxes of inner versus outer, and observer versus observed, that collectively turn on the conundrum of how to understand awareness and subjectivity in a universe that was supposed to be fully describable in objective scientific terms without reference to the mind. The striking paradox is that science tells us both that we're peripheral in the cosmic scheme of things and that we're central to the reality we uncover. Unless we understand how this paradox arises and what it means, we'll never be able to understand science as a human activity and we'll keep defaulting to a view of nature as something to gain mastery over.

Each of the cases just mentioned—cosmology and the origin of the universe, quantum physics and the nature of matter, biology and the nature of life, cognitive neuroscience and the nature of consciousness—represents more than an individual scientific field. Collectively they represent our culture's grand scientific narratives about the origin and structure of the universe and the nature of life and the mind. They underpin the ongoing project of a global scientific civilization. They constitute a modern form of mythos: they are the stories that orient us and structure our understanding of the world. For these reasons, the paradoxes these fields face are more than mere intellectual or theoretical puzzles. They signal the larger unreconciled perspectives of the knower and the known, mind and nature, subjectivity and objectivity, whose fracture menaces our project of civilization altogether. Our present-day technologies, which drive us ever closer to existential threats, concretize this split by treating everything—including, paradoxically, awareness and knowing themselves—as an objectifiable, informational quantity or resource. It's precisely this split—the divorce between knower and known and the suppression of the knower in favor of the known—that constitutes our meaning crisis. The climate emergency, which arises from our treating nature as just a resource for our use, is the most pronounced and catastrophic manifestation of our crisis.

In short, although we have created the most powerful and successful form of objective knowledge of all time, we lack a comparable understanding of ourselves as knowers. We have the best maps we've ever made, but we've forgotten to take account of the map makers. Unless we change how we navigate, we're bound to head deeper into peril and confusion.

Regrettably, the three best-known responses to the crisis of meaning in our scientific worldview are all dead ends.

First, scientific triumphalism doubles down on the absolute supremacy of science. It holds that no question or problem is beyond the reach of scientific discourse. It advertises itself as the direct heir of the Enlightenment.

But it simplifies and distorts Enlightenment thinkers who were often skeptical about progress and who had subtle and sophisticated views about the limits of science. Triumphalism's conception of science remains narrow and outmoded. It leans heavily on problematic versions of reductionism—the idea that complex phenomena can always be exhaustively explained in terms of simpler phenomena—and crude forms of realism—the idea that science provides a literally true account of how reality is in itself apart from our cognitive interactions with it. Its view of objectivity rests on an often unacknowledged metaphysics of a perfectly knowable, definite reality existing "out there," independent of our minds and actions. It often denies the value of philosophy and holds that more of the same narrow and outmoded thinking will show us the way forward. As a result, theoretical models become ever more contrived and distant from empirical data, while experimental resources are applied to low-risk research projects that eschew more fundamental questions. Like Victorian-era spiritualism and pining for ghosts, scientific triumphalism looks backward to the fantasized spirit of a long-dead age and cannot hope to provide a path forward through the monumental challenges that science and civilization face in the twenty-first century.

Science denial on the right and so-called postmodernism on the left represent a second response. These movements reject science. They particularly reject its capacity to establish truths about the world that can be used as a basis for further knowledge and wise policies and actions. Worse still, they provide an opportunity for certain groups to manipulate how facts are interpreted for their own selfish and ideological ends, thereby spreading intentional disinformation in the form of so-called alternative facts and alternative truths. Although the motives of these two movements undeniably differ, both undermine the values of modern society on which they themselves depend, offering nothing but skepticism and negativism, or willful disinformation, in return.

Finally, the new age movement uses fringe science or pseudoscience to justify wishful thinking. Although this movement has had less impact than the other ones, it has muddied the waters for those of us who look for new perspectives on the scientific endeavor. Its uncritical embrace of various misrepresentations of Asian or Indigenous worldviews takes reductionistic science as the norm of all science, and thereby fails to understand the scientific ideas and practices of these other cultures.[1] As a result, constructive dialogue with other cultural traditions about their epistemic practices becomes rare, if not impossible.

Given that these three responses fail to address the crisis of meaning in our scientific worldview, how can we find our way forward? First and foremost, we need to know where the crisis comes from. Our goal is to identify the source of the crisis, offer clues to a new path forward, and present a new perspective on some of the biggest issues science faces today. These issues include time and cosmology, quantum physics and its measurement problem, the nature of life and sentience, how the mind works and its relation to AI, the nature of consciousness, and, finally, climate change and Earth's entry into the new human-shaped epoch called the Anthropocene. The range of topics touched by our new perspective is indeed broad, and the vista is equally expansive. We believe that our perspective can help transform and revive our cherished scientific culture as it faces its greatest challenges while reshaping our worldview for a sustainable project of civilization.

We call the source of the meaning crisis the Blind Spot. At the heart of science lies something we do not see that makes science possible, just as the blind spot lies at the heart of our visual field and makes seeing possible. In the visual blind spot sits the optic nerve; in the scientific blind spot sits direct experience—that by which anything appears, shows up, or becomes available to us. It is a precondition of observation, investigation, exploration, measurement, and justification. Things appear and become available thanks to our bodies and their feeling and perceiving capacities. Direct experience is bodily experience. "The body is the vehicle of being in the world," says French philosopher Maurice Merleau-Ponty, but as we will see, firsthand bodily experience lies hidden in the Blind Spot.[2]

The Parable of Temperature

For a tangible illustration of what we mean by the "Blind Spot," consider an idea as familiar as temperature. We take for granted that temperature is an objective property of the world: it's "out there," independent of us. When we're kids in school, we learn that water freezes at 0 degrees and boils at 100 degrees Celsius. We naturally translate from daily temperature reports in Fahrenheit or Celsius to expectations of how warm or cold we'll feel and what clothes we'll need to wear to go outside. But to distill our now familiar idea of temperature from our bodily sensations of hot and cold took a huge and difficult scientific effort lasting several centuries. Today we view temperature as an objective property of the world, but we've forgotten how the concept of temperature as a physical quantity—the degree or intensity of heat present in an object—derives from the direct experience of the world through our bodies. We have lost sight of the lived experience that underpins the scientific concept and think that the concept refers to something more fundamental than our bodily sensations. This way of thinking is an instance of the Blind Spot.

A little history of science is helpful here. The starting point for creating thermometry (the measurement of temperature) was the bodily sensations of hot and cold. Scientists had to assume that our bodily experience is valid and could be communicated to others; otherwise they would have had no basis for building scientific knowledge. They noticed that sensations of hot and cold correlate with changes in the volume of fluids (liquids expand with heat), and they used sealed glass tubes filled partway with liquids as measuring devices for ordering experiences ("phenomena") as hotter versus colder. These tools enabled them to determine that certain phenomena, like the boiling point of water, were constant enough that they could be used as fixed points for building thermometers with numerical scales.[3]

But once they had invented thermometers, and therefore the concept of temperature, scientists discovered that the boiling and freezing points of water weren't as precisely fixed in the natural world as they initially thought. High on a mountain, for example, water boils at a lower temperature than at sea level. This discovery meant that scientists had to intervene and control the context of their measurements as much as possible by manufacturing true fixed points in highly artificial and controlled settings. This effort required building special places for sequestering the phenomena they were investigating. The aspiring practitioners of thermometry had to build what philosopher of science Robert Crease calls "the workshop," the communal scientific infrastructure required for creating new precise experiences along with the tools for manipulating, investigating, and communicating them.[4]

Once scientists had created the workshop, they used its tools to redefine phenomena in ways increasingly removed from direct experience. The invention of thermometry happened in the absence of any established theory of temperature. But once the ability to measure temperature had been

developed, nineteenth-century scientists took the next step and began for-
mulating the abstract theory known as classical thermodynamics. Then, as
if pulling themselves up by their own bootstraps, they used thermodynamics to define temperature even more abstractly. Now, temperature could
be defined without referring to the properties of any particular substance.
Thermodynamics even allowed for a definition of something physically
impossible—absolute zero, the ideal limit of temperature at which a thermodynamic system has no energy. Later in the nineteenth century, when
physicists devised statistical mechanics, another step in abstraction was
taken as thermodynamic temperature was defined in microphysical terms
as the average motion of molecules or atoms.

The Blind Spot arrives when we think that thermodynamic temperature is more fundamental than the bodily experience of hot and cold. This
happens when we get so caught up in the ascending spiral of abstraction
and idealization that we lose sight of the concrete, bodily experiences that
anchor the abstractions and remain necessary for them to be meaningful.
The advance and success of science convinced us to downplay experience
and give pride of place to mathematical physics. From the perspective of
that scientific worldview, the abstract, mathematically expressed concepts
of space, time, and motion in physics are truly fundamental, whereas our
concrete bodily experiences are derivative, and indeed are often relegated
to the status of an illusion, a phantom of the computations happening in
our brains.

This way of looking at things gives rise to a whole new problem that
expands the Blind Spot: once we eliminate the qualitative character of
bodily experience from the inventory of objective reality, how are we to
account for the sensations of felt qualities like hot and cold? This is the
familiar mind-body problem, which today goes by the name of "the hard
problem of consciousness" or the "explanatory gap" between mental and
physical phenomena.

The downplaying of our direct experience of the perceptual world while
elevating mathematical abstractions as what's truly real is a fundamental
mistake. When we focus just on thermodynamic temperature as an objective microphysical quantity and view it as more fundamental than our perceptual world, we fail to see the inescapable richness of experience lying
behind and supporting the scientific concept of temperature. Concrete
experience always overflows abstract and idealized scientific representations

of phenomena. There is always more to experience than scientific descriptions can corral. Even the "objective observers" privileged by the scientific worldview over real human beings are themselves abstractions. The failure to see direct experience as the irreducible wellspring of knowledge is precisely the Blind Spot.

The tragedy the Blind Spot forces on us is the loss of what's essential to human knowledge—our lived experience. The universe and the scientist who seeks to know it become lifeless abstractions. Triumphalist science is actually humanless, even if it springs from our human experience of the world. As we will see, this disconnection between science and experience, the essence of the Blind Spot, lies at the heart of the many challenges and dead ends science currently faces in thinking about matter, time, life, and the mind.

Our purpose in this book is to expose the Blind Spot and offer some direction that might serve as alternatives to its incomplete and limited vision of science. Scientific knowledge isn't a window onto a disembodied, God's-eye perspective. It doesn't grant us access to a perfectly knowable, timeless objective reality, a "view from nowhere," in philosopher Thomas Nagel's well-known phrase.[5] Instead, all science is always our science, profoundly and irreducibly human, an expression of how we experience and interact with the world. But our science is also always the world's science, an expression of how the world interacts with us. Science strives to be a self-correcting narrative. A successful scientific narrative is made from the world and our experience of it evolving together.

Plumbing the Depths of Direct Experience

It's no coincidence that while science was ascending the spiral of mathematical abstraction and idealization in the nineteenth and twentieth centuries, a movement to plumb the hidden depths of direct experience was occurring in literature and philosophy. Writers such as Emily Dickinson, William Faulkner, James Joyce, Marcel Proust, and Virginia Woolf depicted the subjective stream of thought and feeling, while philosophers such as Henri Bergson, William James, Edmund Husserl, Susanne Langer, Maurice Merleau-Ponty, Kitarō Nishida, and Alfred North Whitehead labored to uncover the primacy of direct experience in knowledge.

A critical moment when these cultural movements intersected was the famous encounter between Henri Bergson and Albert Einstein in Paris on

April 6, 1922. As we discuss later, they debated the nature of time, with Einstein insisting that only measurable physical time exists and Bergson arguing that clock time lacks meaning apart from the direct experience of duration. Historian of science Jimena Canales, in *The Physicist and the Philosopher*, takes their confrontation as emblematic of a growing cultural rift between science and philosophy in the twentieth century.[6] That rift remains in place today, despite many friendly collaborations between physicists and philosophers (such as this book).

Uncovering the Blind Spot can help to repair this rift and the larger split between science and lived experience. But beyond uncovering the Blind Spot, we also need to plumb the depths of the experience it hides.

Drawing from some of the philosophers just mentioned, we will argue that direct experience lies at the heart of the Blind Spot. Direct experience precedes the separation of knower and known, observer and observed. At its core is sheer awareness, the feeling of being. It's with us when we wake up every morning and go to sleep each night. It's easy to overlook because it's so close and familiar. We habitually attend to things instead of noticing awareness itself. We thereby miss a crucial precondition of knowing, for without awareness, nothing can show up and become an object of knowledge.[7]

Philosophers have offered various conceptions of direct experience. William James, the father of American psychology and one of the most influential American philosophers of the nineteenth century, emphasized "pure experience," which he described as "the original flux of life before reflexion has categorized it."[8] A little earlier, Bergson wrote about the experience of "duration," the immediate conscious intuition of passage or flow. The twentieth-century Japanese philosopher Kitarō Nishida drew from Bergson and James but revised their ideas in light of Buddhist philosophy and his experience of Zen meditation practice.[9] Nishida described pure experience as direct experience unmediated by the division between subject and object. Other philosophers have used words like *intuition, feeling,* and the *phenomenal field* to get at this kind of immediate experience or mode of presence. Nishida, in his later writings, used the term *action-intuition* to emphasize that direct experience is not passive and disembodied; to be aware is already to act with our bodies.[10] For example, when you move your eyes, the focus of your awareness shifts. As Nishida writes, "When we think we have perceived at a glance the entirety of a thing, careful investigation will reveal that attention shifted automatically through eye movement, enabling us to know the whole."[11] Direct experience isn't simple and instantaneous; it's complex and has durational rhythms. Crucially, it's prior to explicit knowledge. Knowing presupposes experiencing, and you can't derive experience just from episodes of knowing. Your being is always more than what you know.

In this book, we'll meet some of these philosophers who struggled to articulate direct experience and recover its primacy and who worked to keep our understanding of nature from bifurcating into subject and object, mind and body. Our central concern will always be the dependence of science on experience, a dependence that is far richer and more complex than the obvious dependence of science on observers and experiments. The problem, and the source of the crisis of meaning in our worldview, is that we've become so captivated by the spectacular success of science that we've forgotten that direct experience is science's essential source and constant support.

In plumbing the depths of experience, we have no intention of downplaying the success and value of science. We reject science denial, but we also reject scientific triumphalism. Our quarrel is with a particular, misguided conception of science, one that has come to be built into our present scientific worldview but isn't an essential part of science. This misguided conception, which we delineate in chapter 1, is essentially a philosophy of science based on certain metaphysical assumptions about nature and human knowledge. We argue that science doesn't require this philosophy, and that given its failures, we should jettison it and move on.

In this way, we call for a balanced perspective, where we recognize both the success of science and the problems science has helped to create. The spectacular success of science quickly gave us new vaccines for the worldwide coronavirus pandemic, but science also gave us the conditions of rapid international travel and widespread environmental destruction that made the pandemic possible, with worse future pandemics more likely. Science has helped to create the climate crisis. "The fire next time" is already here. We need another way to understand and practice science, one that doesn't lead to our world burning and that can help to put out the fire we've already started. In short, we need a new kind of scientific worldview. Our starting point is to recover the deep connection between science and human experience that was lost in the Blind Spot.

We began by saying that we write this book with a sense of urgency. The Blind Spot has limited us to a worldview that misunderstands science and impoverishes the living world and our experience. To uncover the Blind Spot and reveal what it hides is to wake up from the delusion of absolute knowledge. It is to embrace the hope that we can create a new scientific worldview, one in which we see ourselves both as an expression of nature and a source of nature's self-understanding. As we describe later, we are caught up in a strange loop in which it is impossible to separate ourselves as knowers from the reality we seek to know. We need nothing less than a science nourished by this sensibility for humanity to flourish in the new millennium.[12]

I How Did We Get Here? A Guide for the Perplexed

1 The Surreptitious Substitution: Philosophical Origins of the Blind Spot

Humanity in Crisis

"We find ourselves in the greatest danger of drowning in the skeptical deluge and thereby losing our hold on our own truth."[1] Edmund Husserl, the twentieth-century German mathematician and philosopher, wrote these words a few years before the outbreak of World War II. They are from his final book, *The Crisis of European Sciences and Transcendental Phenomenology*, first given as lectures in Prague in 1935. Husserl founded the influential movement of phenomenology, which takes experience as its central focus. Born into a Jewish family, he was removed from his university position for being "non-Aryan" when Hitler came to power in 1933. Isolated and discriminated against, Husserl died in 1938, just months before the start of World War II.

Husserl believed that "Western," particularly European, civilization had lost its way. He traced the deep roots of the "crisis of European humanity" to a failure of reason and a fundamental misunderstanding of the meaning of modern science. The confusion was centuries in the making. Science itself, the actual practice of scientists, was not in crisis. On the contrary, science was tremendously successful. Instead, the crisis arose from the meaning that had become attached to science. A particular worldview had been grafted onto science, the worldview we are calling the Blind Spot. The dominant philosophical conception of science led to elevating mathematical abstractions as what is truly real and to devaluing the world of immediate experience, which Husserl called the "life-world." Modern humanity had lost sight of the fact that reality and meaning are far richer than they are represented as being in the dominant materialistic philosophy attached to science. That philosophy had led to a "disenchantment of the world," to use German sociologist Max Weber's term. (Weber was Husserl's contemporary.) In turn, the absorption of the disenchanted world into culture, economics, and politics provoked a backlash of irrational and fanatical efforts to reenchant the world, epitomized by the genocidal Nazi promotion of a racially defined German homeland.

Husserl's rhetoric in *The Crisis* has been dismissed as grandiose, but in today's age of fake news, disinformation, science denial, racism, insurrection, invasion, and the rise of authoritarianism, we might think again. We don't have to accept his ethnocentric narrative about the "telos of European humanity," or agree that his version of phenomenology is the solution to the crisis, to think that he put his finger on a deep problem endemic to our scientific culture. The polarization between science triumphalism and science denial, combined with the existential threat to the life-world from human-caused climate change, indicates that we have exacerbated the problem in this century. Husserl's crisis is still our crisis.

It would behoove us, therefore, to consider Husserl's diagnosis, particularly his analysis of how we misunderstand science and the life-world when we promote abstract mathematical entities to the rank of what is truly real while downgrading our concrete perceptual experience. Alfred North Whitehead, another early twentieth-century mathematician and philosopher, also trenchantly criticized what we are calling the Blind Spot. Their writings, combined with recent ideas from the philosophy of science, will enable us to identify the key elements of the Blind Spot worldview and to see why they are unsound. Our first step is to describe the general philosophical contours of this worldview.

The Air We Breathe

The Blind Spot worldview is like the air—invisible but all around us. We're given simple versions of it in high school science classes, and we find it as an unspoken background assumption in science documentaries. If you pursue a career in science, it often lies like an invisible map marking your journey through introductory classes in physics, chemistry, and biology. Although there have been sophisticated philosophical articulations of the ideas that make up what we call the Blind Spot worldview, for most people, including most scientists, it's so pervasive that it doesn't seem like philosophy at all. Rather, people think it's just "what science says."

In truth, however, it's not what science says. Instead, it's an optional metaphysics attached to, but separable from, the actual practice of science. There are other and better ways to understand the relationship between science and the world, as we argue here.

Although the Blind Spot arises from and expresses a particular philosophical viewpoint, it isn't a theory. Instead, it's a broad perspective that encompasses many different problematic theories and ideas. These include opposed positions on various scientific and philosophical issues, so the ideas and theories it embraces don't always have to be consistent with each other. For example, with regard to the problem of how the mind relates to the body (the mind-body problem), the Blind Spot includes both materialism or physicalism, according to which states of consciousness are strictly identical to (one and the same thing as) brain states, and dualism, according to which consciousness is a special mental property irreducible to the brain. In addition, in the philosophy of science, the Blind Spot can include realism, the position that scientific theories aim to give a true account of a mind-independent reality, or instrumentalism, the view that scientific theories are mainly tools for making successful predictions about observations. Nevertheless, despite these differences, it's possible to summarize what the Blind Spot involves in general philosophical terms. We list the main ideas below.

1. **The bifurcation of nature.** *Color is an illusion, not part of the real world.* This idea isn't based on the actual practice of physicists when they measure light, or construct particle or wave models of light, but rather on physicists' theoretical reflections, which is to say, their philosophical thoughts. They divide nature into what exists externally and objectively, and what is mere subjective appearance or exists just in the perceiver's mind. Light waves, with mathematical properties such as amplitude, frequency, and length, really exist outside the perceiver in nature, whereas color is said to be just a subjective appearance or perceptual illusion. The illusion of color is ultimately to be explained (or explained away) in terms of what really exists, so that red, for example, is just a subjective sensation or perceptual illusion caused by electromagnetic radiation of particular wavelengths. Similarly, temperature, defined as the average kinetic energy of atoms or molecules, really exists; hot and cold are mere sensory appearances. Or, if we take the idea to the limit, particles and

forces are fundamentally real; visible and tangible objects are illusory perceptual constructs. Whitehead calls this way of thinking "the bifurcation of nature," because it splits nature into external reality and subjective appearance. We will discuss the bifurcation of nature extensively.

2. **Reductionism.** *Elementary particles are the fundamental building blocks of matter and everything in the universe reduces to them.* Reductionism is a complex constellation of ideas. One idea is *smallism*, the idea that small things and their properties are more fundamental than the large things they constitute. People are more fundamental than the social groups they compose, cells are more fundamental than the people they compose, molecules are more fundamental than the cells they compose, atoms are more fundamental than the molecules they compose, and, finally, elementary particles are more fundamental than anything else. Smallism implies that the real action of nature happens at the smallest microphysical scale, because that's where the most basic causal processes are thought to occur. Smallism belongs to the province of what philosophers call ontology, the theory of what kinds of things exist and what their relations to each other are. Another part of reductionism belongs to the province of epistemology, the theory of knowledge and explanation. This part is the idea that the main preferred method for explaining a system is *microreduction*, which consists in breaking a system up into its elements and explaining the properties of the whole in terms of the properties of its parts. Smallism and microreduction imply that elementary particle physics is the preeminent science, because it's the only science that remains once you progressively reduce statements about large things or systems in sociology, psychology, biology, and chemistry to statements about the smallest things in fundamental physics. Reductionism is summed up in the quip, "Biologists defer to chemists, who defer to physicists, who defer to mathematicians, who defer to God."

3. **Objectivism.** *Science strives to attain a God's-eye view of reality as a whole.* This is the idea that science, particularly fundamental physics, aims to provide access to reality apart from any human perspective. Science discovers truths about observable and unobservable aspects of mind-independent reality. Fundamental physical entities are mind-independent, real objects and have their essential properties independent of any observation.

4. **Physicalism.** *Everything that exists is physical.* If you make a list of everything that exists in the universe, everything on the list will be completely physical in its nature and constitution. And, if you make a list of all the physical facts in the universe, you will ipso facto have determined every state of affairs in the universe, including every chemical, biological, psychological, social, and cultural state of affairs. So physical facts exhaust reality. *Materialism* is the older term for this idea, but philosophers today prefer *physicalism*, because physics has shown that not everything is material in the classical sense of being an inert substance with the property of extension. For example, fields and forces are physical but not material. Physicalism is first and foremost a general metaphysical thesis, not a scientific one. It's not a thesis that belongs to any theory of physics. It's a philosophical interpretation of physics and science in general. Today the main obstacle to physicalism is considered to be how to account for the mind, particularly consciousness, within a physicalist framework.

5. **Reification of mathematical entities.** *Mathematics is the language of nature.* This is the idea that the real structural properties of the universe are its mathematizable properties. The mathematical entities in scientific models, laws, and theories exist "out there" apart from us. They alone constitute the true structure of the universe. As Galileo wrote, echoing an older idea going back to Plato and Pythagoras, "The universe . . . is written in the language of mathematics."[2]

6. **Experience is epiphenomenal.** *Consciousness is the brain's user illusion.* A user illusion is a visual image, such as the image of a desktop on a computer screen, created for the convenience of the person who uses the computer. Similarly, conscious experiences are said to be representations generated by the brain for the convenience of its operation in controlling the body's interactions with the world. Conscious experiences are no more real than the desktop icons on your computer. Experiences result from computations going on in your brain but play no significant role in those computations. Subjectivity, the experience of being, is a derivative effect of physical events in the brain and has a mostly insignificant influence on those events. The impression that consciousness is efficacious is largely an illusion.

This list isn't meant to be exhaustive or definitive. We aren't saying that these ideas are individually necessary and jointly sufficient for the Blind

Spot. Scientists and philosophers have combined various aspects or sub-sets of these ideas in various ways over the centuries. Our aim, again, is to identify a cluster of ideas that shape our present scientific worldview and that many people, including many scientists, now mostly take for granted.

Although we have presented these ideas in philosophical terms, the Blind Spot is much more than a perspective to be debated among academic philosophers. Given the far-reaching authority and power of science, any position successfully claiming to speak for science will have tremendous social and environmental influence. The blind spot worldview has great impact as a social force that constrains people to think in certain prescribed ways about how science works, how human life fits into the biosphere of our planet Earth, and how the human mind relates to the cosmos. Further-more, the development and deployment of science and its offspring tech-nologies over the past four centuries have been inseparable from enhancing economic and military power, whether in capitalist or socialist countries. As we discuss later, blind spot conceptions of nature, energy, and informa-tion have framed how we think about natural resources, energy production, information technologies, and artificial intelligence. For these reasons, it's important to remember that the Blind Spot is like the air we breathe: it's a culturally ubiquitous mind-set and not a constellation of abstruse philo-sophical ideas.

The Surreptitious Substitution

We turn now to Husserl's critique of what we are calling the Blind Spot. Husserl's training as a mathematician gave him insight into how we deploy abstractions in the sciences, particularly in mathematical physics. He called the elevation of mathematical constructs like that of temperature to the status of fundamental reality "the surreptitious substitution." For Husserl, this substitution is a fundamental mistake.

In the development of the modern scientific worldview, which Husserl regards as beginning with Galileo, the abstract and idealized representation of nature in mathematical physics is covertly substituted for the concrete real world, the world that we perceive. The perceptual world is demoted to the status of mere subjective appearance, while the universe of mathemati-cal physics is promoted to the status of objective reality. Thus, according to this way of thinking, temperature or the average kinetic energy of atoms or molecules is what's objectively real, but the feelings of hot and cold are mere subjective appearances.

Husserl argues that the surreptitious substitution is unjustified because it's based on a fundamental misunderstanding of mathematical physics. Physical laws specify, in mathematical terms, how things behave in idealized situations. For example, Galileo's law of free-falling bodies states that in the idealized case of the absence of air resistance, all bodies fall with the same acceleration independent of their mass. Galileo showed mathematically and experimentally that the distance a free-falling body travels is directly proportional to the square of the time it takes to fall. Similarly, the ideal gas law, also known as the general gas equation, states the relation between the pressure, volume, and temperature of an ideal gas. In this equation, the molecules do not attract or repel each other or take up any volume. The molecules are instead represented as geometrical points called "ideal structureless particles."

We can see from these examples that the laws of mathematical physics refer to idealized objects and their properties—free-falling bodies, frictionless planes, hypothetical ideal gases, perfectly elastic collisions, and so on. These idealized objects and properties aren't physically real. They don't actually exist in space and time, and they don't participate in causal interactions. So they do not and cannot constitute the real world of nature. They're fictional entities that we use as tools. They're conceptual instruments necessary for us to formulate exact mathematical statements that we can apply to the real world through a series of increasingly accurate approximations. This is how we gain predictive knowledge of things and control over them.

Mathematical idealization and approximation constitute a method for knowing how things will behave under various conditions. But the method doesn't tell us what things are and why they behave as they do. Hence, to think that the idealized laws of mathematical physics describe the inherent being of nature is fundamentally mistaken. To think this way is to confuse the map—an idealized and limited representation of the terrain—with the territory. As Husserl says, to think this way is to "take for *true being* what is actually a *method*."[3] This is the surreptitious substitution—the substitution of a tool to describe phenomena for nature itself, or confusing an instrument of prediction for how things are in themselves. It's a kind of category mistake.

In philosophy of science terms, Husserl can be described as an *instrumentalist*, or more generally an *anti-realist*, about scientific laws (but not necessarily about scientific entities, such as electrons, as we discuss below).[4] According to his view, laws are precise instruments of prediction, and the idealized objects of the laws have an ideal, mathematical status, not a real, physical existence.

Nancy Cartwright, an American and British philosopher of science, argues for a similar anti-realist conception of scientific laws in her classic book *How the Laws of Physics Lie*.[5] According to Cartwright, mathematical physical laws don't describe reality; they describe idealized objects in models. We need these models so we can apply the abstract theories of mathematical physics to the real world. For example, in classical mechanics, the concept of force is abstract and needs to be replaced with a specific force such as gravity to describe the attraction between massive bodies. This replacement is made using more simplified and idealized models—for example, that of two spherical bodies (approximated by two point-like objects with masses) in an otherwise empty space. This simplification makes the equations of motion manageable. Scientific description consists in moving from abstract, theoretical concepts to concrete, real-world applications via idealized models. Cartwright calls these "interpretative models."[6]

Galileo's model of a frictionless plane, the Bohr model of the atom with a dense nucleus surrounded by electrons in quantized orbits, evolutionary-biological models of totally isolated populations—these are idealized representations that exist in the minds of scientists. They are not concrete realities in nature. We should not surreptitiously substitute abstract mental representations for concrete reality, the map for the territory. As we discuss next, Husserl also thought that we should not surreptitiously substitute the scientific workshop for the world.

The Scientific Workshop

Today we take for granted that science requires specialized laboratories that are housed in universities and research institutes, supported by government and private funding agencies. These labs collaborate internationally to foster technological innovation and train the next generation of scientists. But this global scientific infrastructure, the scientific workshop in Robert Crease's terminology, is little more than a few centuries old.[7] It's a very recent achievement in the history of human civilization, let alone in human history altogether.

The idea of the workshop goes back to the sixteenth century. Francis Bacon, the English philosopher and statesman who lived from 1561 to 1626, was the first person to envision the scientific workshop.[8] He had the idea of creating dedicated facilities in which to investigate and control nature systematically using experimental methods and specialized tools. He bequeathed to us the idea of experimental science as a collective enterprise housed in research institutes that work to benefit humanity.

Bacon is a difficult and ambiguous figure. Historian Carolyn Merchant, in her classic work *The Death of Nature: Women, Ecology, and the Scientific Revolution*, argues that although Bacon advanced egalitarianism by establishing the inductive method by which anyone, in principle, can verify for themselves the truths of science, he also helped to undermine communal agrarian society in favor of "an emerging market economy that tended to widen the gap between upper and lower social classes by concentrating more wealth in the hands of merchants, clothiers, entrepreneurial adventurers, and yeoman farmers through the exploitation and alteration of nature for the sake of progress."[9] In other words, Bacon's conception of an inductively based scientific enterprise conducted in the workshop was thoroughly in the service of the emerging capitalist world system and its conception of nature as essentially a resource for humanity. Bacon also used female imagery to describe nature and its subjugation by science. As Merchant explains, "Nature cast in the female gender, when stripped of activity and rendered passive, could be dominated by science, technology, and capitalist production. . . . Francis Bacon advocated extracting nature's secrets from 'her' bosom through science and technology. The subjugation of nature as female . . . was thus integral to the scientific method as power over nature."[10] It's precisely this modern conception of nature as something to be subjugated by science and technology that has led to human-caused global warming and the climate crisis.

In Bacon we see the beginning of a tendency in modern scientific thought to which Husserl would strenuously object. We're referring not to the oppressive gender imagery—it's unlikely Husserl would have been sensitive to this—but instead to the tendency to view nature outside the workshop entirely in terms of concepts and procedures applied to nature inside the workshop. Inside the workshop we sequester phenomena, protect them from outside influences, subject them to our specialized devices, and thereby manufacture new phenomena. The rest of the life-world, on which the workshop depends for its purpose and meaning, resides outside the workshop. Husserl affirms that when we manipulate phenomena in the workshop, we enlarge our understanding of reality and acquire new means of control over things. But he objects to making the workshop's conceptual and experimental tools the final arbiters of what is real. In Crease's words, Husserl rejects the idea that "reality is what turns up in the scientific workshop"—that reality can be equated with or be defined in terms of objective measurements in the workshop.[11] That idea is a version of the surreptitious substitution: it substitutes the workshop for the world.

You might be skeptical about this line of thought. Isn't basic science, especially fundamental physics, all-embracing? Why can't we just transfer scientific theories from the workshop to the world?

The World in the Workshop and the Workshop in the World

As Cartwright discusses, the predictive models of physics work mostly inside walls—the walls of a laboratory, a particle detector, a large thermos, a battery casing, and so on.[12] In other words, the models work in places that we can control and shield from outside influences and where we can precisely arrange the conditions to fit the models. Sometimes the models also work outside of walls, in cases where local patches of nature sufficiently resemble one of our workshop models. Often these cases involve things we build to fit our models, such as airplanes and rockets, which we engineer so that they conform to models in classical mechanics, or computers, where we use quantum mechanics to design semiconductor elements that precisely implement the computations of digital logic.

We make a surreptitious substitution, however, when we suppose that such physical or computational models apply to the rest of nature, including the human life-world, beyond the range of what we're able to model and successfully predict. For example, we suppose that there must be, in principle, a predictive mechanical model for the action of the wind and water on a lost seagull feather swept along by the breeze and waves at the beach. Or we suppose that there must be, in principle, a predictive computational model for the interactive brain activities of the improvisational players in a jazz quintet. Of course, we allow that we will probably never be able to formulate these models because they are too complex. Nevertheless, we suppose that they are formulable in principle.

We should be wary of this way of thinking, and we will criticize various instances of it in this book. It rests on a loaded and unnecessary metaphysical assumption about what the world is like outside the range of our ability to construct and test predictive models. The assumption is that how things behave in tightly controlled and manufactured environments should be our guide to how things behave in uncontrolled and unfabricated settings. In other words, the assumption is that the regimented facts inside the workshop are exemplary of the world outside the workshop. As Cartwright discusses, this way of thinking amounts to a kind of fundamentalism, because it requires strict adherence to workshop models as literal truths about how the world fundamentally is outside the workshop.[13]

This fundamentalist attitude usually goes hand in hand with the idea that objective measures in the workshop are more valid than direct experience of the world based on our bodily perception. But this objectivist idea is misguided, as Crease shows using the example of surfers (drawn from William Finnegan's memoir, *Barbarian Days*[14]). Objective measures of a wave's height are irrelevant to surfers; what matters to them is how they gauge the size and ferocity of a wave in relation to their bodies and surfing skill, which is to say in relation to their direct experience. How surfers, sailors, and swimmers interact with waves through direct experience is what matters most to them, not objective measures of waves apart from how they're experienced.

The combination of fundamentalism and objectivism about scientific models exemplifies the Blind Spot. It occludes our direct experience of the world outside the scientific workshop. It advertises itself as "what science says," when it's really a philosophical mind-set, not anything established by actual scientific practice. It's based on generalizing from a small number of cases where we do have successful predictive models to a vastly larger number of cases where we do not, and arguably cannot, have this kind of knowledge, because the world outside the workshop is too entangled and complex. As network science increasingly suggests, how things behave, especially in the world outside the workshop, often depends more on the global, intertwining structure of events than on the local arrangement of parts that we can sometimes isolate and shield. It follows that we will go badly wrong, not just in theory but also in practice and social policy, if we treat the world as if it were just a bigger version of what happens inside the workshop.

The Amnesia of Experience

Underlying the surreptitious substitution is a kind of amnesia of experience. We take this idea from Michel Bitbol, a French philosopher of science who draws from Husserl.[15] Bitbol points out that we produce objective knowledge in two main steps. First, we progressively set aside anything in our experience on which we cannot find absolute agreement, such as how things feel or look, or our individual preferences, tastes, and values. In other words, we progressively abstract away from concrete experience. Second, we retain a "structural residue" of experience that we can make into an object of consensus, especially when we refine it in the workshop. Structural residues include classification schemes (taxonomies), models, general propositions, and logical systems. The most abstract kinds of structural residues are mathematical, such as magnitudes. The amnesia occurs when we forget that direct experience is the implicit departure point and constant requirement of this procedure of creating objective knowledge.

Our parable of temperature (from the Introduction) provides an illustration. The departure point is the experience of hot and cold. First, we remove anything on which we cannot definitely agree, such as whether a bowl of water is warm or cold, given that it feels one way to one hand and another way to another hand (to use Locke's example[16]). In this way, we abstract away from particular felt qualities. Second, we progressively distill phenomena, in the form of noteworthy experiences on which we can agree, such as observed correlations between sensations of hot and cold and changes in the volume of fluids (liquids expand with heat), and observed points at which water starts to freeze or boil. We use these relatively stable points to construct thermometers with numerical scales so that we can now describe the phenomena in terms of consensual objective magnitudes. Refining this procedure in tandem with developing physical theories (classical thermodynamics and statistical mechanics) eventually leads to the abstract concept of thermodynamic temperature (the average kinetic energy of particles). This abstract concept is an example of a highly refined structural residue of experience. Bitbol calls the corresponding abstract object (kinetic energy due to the motion of particles) an "invariant structural focus" for experience.

So far, so good. The problem is when experience drops out of the story. We become so captivated by the success of the method that we forget its necessary experiential dimension. This is the amnesia of experience. This is the Blind Spot.

The amnesia of experience eventually leads to a strange and nonsensical idea, prevalent in certain quarters of science and philosophy, that experience can be reduced to one or another of its structural residues. A striking case is the idea that conscious awareness can be reduced to the structural residue of informational or computational processes in the brain. This way of thinking inverts the whole procedure of producing objective knowledge by supposing that an abstract structural residue of experience (such as "information") can explain or ground the concrete being of conscious awareness. Here we encounter the Blind Spot at its most extreme.

Husserl (and Whitehead, as we will see) clearly recognized that this way of thinking is absurd *in principle*. Abstract scientific concepts (thermodynamic temperature, information, computation) spring from concrete experience and therefore cannot explain or ground experience. The abstract can never explain or ground the concrete as a matter of general principle. Rather, the reverse always is and must be the case. Part of the crisis of our scientific culture is that we have allowed ourselves to forget this fundamental truth.

Are Unobservable Objects Real?

So far our discussion has focused on scientific models and laws. We've argued that the ideal objects and properties of scientific models and the idealized regularities of scientific laws aren't physically real, so a surreptitious substitution occurs when we treat them as if they were truly real and the perceptual world were merely apparent. We've also emphasized that most of the phenomena of modern physics are manufactured in the workshop, so another surreptitious substitution happens when we replace the world with the workshop.

Models and laws, however, are hardly all of science; there are also scientific theories. According to the standard textbook conception of science, whereas models and laws describe what happens, theories provide an overarching explanatory framework. For example, Einstein's special theory of relativity postulates that the laws of physics (the statements that describe or predict physical phenomena) are invariant in all inertial frames of reference (frames of reference with no acceleration) and that the speed of light is a universal constant. In biology, Darwin's theory of evolution explains speciation as a result of the process of natural selection. Or take the kinetic theory of gases. It explains how the behavior of atoms or molecules gives rise to the macroscopic regularity known as Boyle's law, which states that the pressure of a gas varies inversely with its volume at constant temperature. A crucial feature of scientific theories is that they postulate entities, such as atoms and electrons, or genes, that causally interact to produce observable behavior but that we can't observe directly. Models use such entities to describe observed data.

Postulating unobservable entities like atoms raises the question of what reality should be accorded to them. We think that postulating certain kinds of unobservable entities and thinking they are physically real can be perfectly legitimate. The problem arises when we lose sight of how direct experience lies behind this kind of scientific procedure. This is another instance of the Blind Spot.

We first need to recognize that although scientific laws and scientific theories may both refer to unobservable objects, the kinds of objects and the reasons that they're unobservable differ in the two cases.[17] Abstract and ideal objects, such as point particles or an ideal gas, aren't possible objects of perception, so the ideal objects to which the laws of mathematical physics refer can't possibly be perceived—by any possible perceiver, not just human beings.

The unobservable entities to which physical theories refer, however, such as particular kinds of molecules, atoms, and subatomic particles, aren't ideal; they're supposed to be real. They're spatiotemporal and participate in causal interactions. Although we can't see molecules and atoms with our naked eyes, we can image them thanks to advances in microscopy. In the twentieth century, cloud chambers and bubble chambers were used to detect electrically charged, subatomic particles. Here physical theory was used to interpret what we see with our eyes—condensation trails of water droplets—as indicating the presence of new kinds of entities, namely, specific kinds of subatomic particles, such as muons. In other words, using physical theory, we were able to learn to see the condensation trails as signs or tracks of these particles. Finally, although molecules, atoms, and subatomic particles are imperceptible to human beings, it doesn't follow that they're imperceptible to any logically or conceptually possible perceiver. There's nothing logically or conceptually contradictory in the supposition that there could be a kind of perceiver, such as an alien with special sense organs, capable of perceiving them. By contrast, it's logically or conceptually impossible for ideal objects like a perfect spherical charge to be perceived because they aren't real; they aren't spatiotemporal, and they don't participate in causal interactions, so there's no way for them to be given to sense perception.

The next step is to recognize that the scientific theories and experiments that lead us to postulate unobservable entities don't take us outside experience; they enlarge and enrich it. For example, twentieth-century physicists used cloud chambers to discover the positron, muon, and kaon particles. Although we can't directly see these elementary particles, physicists were able to see signs of their presence in the form of the curvature and direction of trails of small droplets. Given the theory of particle physics and the precise observations made in the highly manufactured and controlled conditions of the cloud chamber—a major tool of the scientific workshop of particle physicists from the 1920s to 1950s—positing the existence of those particles was the best explanation of the data, including especially what was directly observable in the cloud chamber.

This form of inference is usually called "inference to the best explanation," but we prefer Cartwright's term, "inference to the most probable cause."[18] We rely on this kind of inference all the time in everyday life. For example, you enter the kitchen in the morning to find a mess of dirty dishes and most of the groceries you bought yesterday gone from the fridge. You immediately conclude that your teenage kids made themselves a midnight meal and were too lazy to clean up. This is the most probable cause, given many other things you know and have observed; it's certainly more probable than that your house was broken into by hungry people who cooked a late-night dinner and left without cleaning up.

Although inference to the most probable cause in the scientific workshop is more rigorous and constrained than this example, it's essentially the same kind of reasoning procedure. This kind of inference is central to how we experience the world and find our way around in it. So even though physicists in the workshop use inference to the most probable cause to infer the presence of unobservable entities that go beyond the range of human perception, they're still operating entirely within the sphere of human experience.

Just as important, the content of the observations in the workshop is enriched when the existence of these particles is established using inference to the most probable cause. The observations now count as what we can call "observations of perceptual signs," as when we see the tracks of some animal in the snow. Indeed, physicists call the variously shaped trails formed by tiny droplets in a cloud chamber the "tracks" of electrically charged particles. These tracks belong to the real world of human perception. Our perception of them is a kind of "seeing as": we see the droplet trails *as* tracks. We have good grounds for seeing them this way and for inferring that they are caused by particles. Just as we can tell that a deer went up a hill by seeing its tracks in the snow, so the physicist can see that the particle is a positron from the shape and orientation of its condensation trail. The shape of the trail functions as an "indicative sign" of the particle, to use Husserl's term. This kind of perception—perceiving something as the sign of something else, "indicative sign consciousness"—is fundamental to how we perceive the world.[19]

Of course, there's a difference between seeing the tracks of a deer and the tracks of a particle. We can follow the deer tracks and eventually see the deer. Following the tracks of the particle will never lead us to see the particle with our naked eye. This fact, however, shouldn't lead us to say that the particle doesn't exist or that it's just a fictional posit in the way that the mathematical objects of idealized physical laws are fictional posits. Although we can't see particles, we can manipulate them, as when physicists spray electrons from an electron gun. As Canadian philosopher of science Ian Hacking wrote forty years ago, "If you can spray them, then they are real."[20] (Or in Cartwright's more cautious emendation, "*When* you can spray them, they are real," since they're sprayed only under precisely controlled conditions in the workshop.[21]) This marks another difference between the unobservable entities of physical theory and the unobservable objects of physical law: the latter are not only imperceptible in principle; they're also unmanipulable in principle precisely because they aren't spatiotemporal and can't participate in causal interactions. You can't spray an ideal gas.

We've been arguing that we can have good reasons for postulating the existence of unobservable entities, but that these reasons depend essentially on direct experience—our experience of perception (seeing the tracks of particles) and our experience of action (spraying electrons). Thinking that elementary particles and fields of force are real, however, does not imply that the world we perceive is somehow less real than these entities. This simply does not follow logically. Taking this further step of demoting the reality of the perceptual world and replacing it with the entities of particle physics would be a case of the surreptitious substitution.

The problem with the surreptitious substitution is that it undermines the ground on which science must stand. Science must presuppose the concrete reality of the perceptual world, especially the macroscopic things that we perceive and manipulate in the scientific workshop. The scientific workshop functions only in the larger setting of the life-world, the immediately given world of concrete things, situations, and human communities with their values and projects. Science is essentially a project within the life-world; it can enlarge and enrich our conception of reality, but it can't step outside the life-world.

In addition, calling into question the validity of our human mode of perception would make science lose its epistemic footing. For example, inference to the most probable cause requires that the states of affairs on which it's based be perceivable. The kitchen really is a mess, so you conclude your kids didn't clean up after themselves; there really is an observable trail of droplets with a particular shape, so you conclude that the positron exists. Similarly, for something to be an indicative sign of the presence of something else, it must be perceivable, like tracks in the snow or tracks in a cloud chamber. More generally, our concrete bodily experience provides the starting point and final source of evidence for science, so to impugn its validity altogether, to make it an epiphenomenal sideshow, is to abandon the true basis for science.

The Bifurcation of Nature

It's time to bring Alfred North Whitehead into the discussion. Whitehead clearly recognized what we are calling the Blind Spot, but he also charted a new philosophy of science and nature beyond it. He upheld science and the importance of abstraction, but he refused to allow the abstract to usurp the concrete, and he insisted on not bifurcating nature into objective reality versus subjective appearance. Whitehead recognized that science has its source in direct experience, that perception already abstracts from what is available to it through the medium of our living body and that science rigorously extends this abstraction without ever leaving experience behind.

In his 1920 book *The Concept of Nature*, Whitehead begins from the observation that nature is that of which we're aware in sense perception.[22]

Physicists, however, must abstract away from perception to focus on the mathematizable aspects of nature, such as time, space, motion, forces, and particles. So far so good. The "bifurcation of nature" happens whenever this procedure of abstraction is interpreted to mean that nature consists of two different kinds of things, the unperceived and fundamental constituents of reality and what appears subjectively in perception. We bifurcate nature when we divide it into "two systems of reality": "the nature apprehended in awareness and the nature which is the cause of awareness."[23]

The scientific bifurcation of nature arose in the seventeenth century in the form of Galileo's, Descartes', and Locke's distinction between "primary qualities" (size, shape, solidity, motion, and number), which are thought to belong to material entities in themselves, and "secondary qualities" (color, taste, smell, sound, and hot and cold), which are thought to exist only in the mind and to be caused by the primary qualities impinging on the sense organs and creating sensations or mental impressions.

Although our understanding of matter changed dramatically in the nineteenth and twentieth centuries, the bifurcation of nature remained in place. A modern example is twentieth-century physicist Arthur Eddington's "two tables," the ordinary table and the scientific table. The ordinary table has shape, color, and solidity, and it belongs to the commonplace world of perception. The scientific table is mostly empty space pervaded by fields of force. Modern physics assures us, according to Eddington, that the scientific table is the only one that is really there.[24] Herein lies the problem characteristic of the Blind Spot: the demotion of concrete experience and the elevation of abstractions.

Whitehead steadfastly refuses to accept the bifurcation of nature because it violates the premise of natural philosophy (the older name for science) and has unacceptable consequences. The premise of natural philosophy is that "everything perceived is in nature. We may not pick and choose."[25] Nature, for the empiricist tradition from which Whitehead is working, is "that which we observe in perception through the senses."[26] Better yet, sense perception is the touchstone of nature, because nature also includes that which we can rigorously abstract from perception using logic, mathematics, and controlled experimentation. Thus, we infer the existence of atoms, electrons, and other subatomic particles through detailed experiments. But as Whitehead says, the red of the sunset is as much a part of nature as the molecules and light waves by which scientists explain it.[27]

The task of natural philosophy is to analyze how these elements—red, molecules, light waves—are connected in one system of relations. "There is but one nature before us, namely, the nature which is before us in perceptual knowledge."[28] The bifurcation theory, however, a fundamental aspect of the Blind Spot, cuts nature in two, making one part real and the other part apparent, with the real part being the cause of the apparent part. Thus, temperature (the average kinetic energy of particles) and light (part of the electromagnetic spectrum) really exist, whereas hot and cold, and color, are mere appearances. Or the "scientific table" (particles, space pervaded by fields of force) really exists, but the ordinary table is just a perceptual image. This cut separates the world of perception from the rest of nature and leaves no way to put them back together. A hidden physical reality causes the perceptual world, which is lodged in the mind: "Thus there would be two natures, one is the conjecture and the other is the dream."[29] The dream (hot and cold, color, ordinary perceptual objects) is supposed to be explainable in terms of the conjecture (particles, electromagnetic radiation), but this turns out to be impossible because the conjecture presupposes the dream. The postulated hidden physical reality is an abstraction from the concrete elements of nature present to us in sense perception and can't exist on its own apart from those concrete elements.

Whitehead is hardly opposed to abstraction (he's a mathematician, after all). But he is opposed to placing abstractions above concrete experience in a hierarchy of knowledge and mistaking abstractions for concrete realities. Our parable of temperature illustrates the point: kinetic energy due to the motion of atoms and thermodynamic temperature as a measure of atomic motion are abstractions from the concrete reality of hot and cold. Whitehead's point is that thermodynamic temperature is abstract, whereas hot and cold are concrete, and that abstractions presuppose a concrete basis, so we should not think that thermodynamic temperature is more real than hot and cold. As philosopher of science Isabelle Stengers writes in *Thinking with Whitehead*, "Between the most concrete experience and the various abstractions, there is no hierarchy for Whitehead. . . . The question that worries him, as always, is that of abstraction and, more precisely, the lack of resistance characteristic of the modern epoch to the intolerant rule of abstractions that declare everything that escapes them frivolous, insignificant, or sentimental."[30]

In *Science and the Modern World*, published in 1925, Whitehead calls the surreptitious substitution of abstractions for concrete realities "the fallacy of misplaced concreteness."[31] This is "the accidental error of mistaking the abstract for the concrete"—accidental because it's unnecessary for the actual practice of science.[32] The bifurcation of nature exemplifies the fallacy of misplaced concreteness, because it arises from mistaking the mathematical abstractions of physics (average kinetic energy of atoms or molecules) for concrete reality (hot and cold) and proposing to reduce the concrete to the abstract. This reduction nonsensically reverses the dependence relation (the abstract always presupposes the concrete).

Once we bifurcate nature, we cannot help but place the mind outside nature while nonetheless finding ourselves forced to invoke it in a mysterious way to explain what appears in perception. Suppose we say that light waves and brain activity cause us to see red. Given the bifurcation of nature, light waves and brain activity are physically real, whereas red is a subjective appearance. But subjective appearances are mental phenomena. So in saying that light and the brain cause red, we invoke an illicit kind of causality, one that jumps mysteriously from material entities to foreign mental ones. We posit a causal relation that no longer operates within the one domain of nature. Instead, it traverses nature and the foreign domain of the mind. In other words, once the mind is pushed outside nature, which inexorably happens with the bifurcation of nature, any physical-to-mental transaction cannot count as an interaction within nature. Natural science, however, is premised on not referring to anything outside nature in its explanations. Hence, appealing to the mind, even just as the supposed end point of a causal chain, is unacceptable.

As Whitehead observes, from a natural-science perspective, we do not actually explain red when we mention light entering our eyes and the subsequent brain activity.[33] Instead, we presuppose the perceptual presence of red and determine that it's accompanied by the brain's responding to the eye's reception of light. Hence, we should say that the perceptual presence of red in nature interdependently occurs with light entering our eyes and the ensuing brain activity, not that light and brain activity cause the mental occurrence of red. Whereas the second way of speaking bifurcates nature (into real versus apparent, physical versus mental), the first way of speaking does not (there is only one system of interdependent events).

Invoking the mind or "psychic additions," in Whitehead's terms, to explain how physical reality causes the apparent world of perception leaves the mind dangling on its own, unrelated to anything else. When we try to incorporate the mind back into nature, we face an intractable problem, the mind-body problem. Today's version—How is consciousness possible in a purely physical universe?—is an artifact of the bifurcation of nature and the Blind Spot.

Advances in neuroscience expand the problem but do not solve it. Far more than in Whitehead's time, we're now able to delineate many of the physical and neuronal events that occur interdependently with conscious perception. But we're stymied by the so-called explanatory gap between physical nature and conscious experience. Whitehead would tell us that the explanatory gap results from nature's bifurcation and there's no way beyond the gap as long as that split is maintained. Cognitive neuroscience can't bridge the gap because it presupposes the scientist's first-person experience of perception. Stengers puts her finger on the point: "From explanation to explanation, from experimental situation to experimental situation, the 'mind' remains in brackets, in the sense that the characterization of the experimental relation does not include the person for whom it constitutes an explanation. To claim to explain the 'perceived red' is to claim that there will suddenly surge forth, like a rabbit pulled out of a magician's hat, a term that designates not what the scientists have succeeded in perceiving but that which all their achievements presuppose."[34] In other words, although scientists are establishing ever more precise correlations between neuronal activities and reports of conscious perception in their experimental participants, these successes presuppose the consciousness of the scientists themselves as a precondition of the whole scientific enterprise. Thinking that the same methods we use to study correlations between brain activities and reports of conscious experience can be turned back onto consciousness as a precondition of the presence of anything, particularly to science, is like thinking you can pull a rabbit out of an empty magician's hat.

The problem of how to incorporate conscious experience into a physical nature from which it has been deliberately expunged is a badly formulated problem. It's symptomatic of the Blind Spot. Nevertheless, it does not follow that understanding the interdependent relation between brain activity and what we're aware of in experience poses no problem whatsoever. Whitehead recognizes this: "There really is a difficulty to be faced relating within the same system of entities the redness of the fire with the agitation of the molecules."[35] What he refuses to do in facing this difficulty is to allow nature to bifurcate: "So far as reality is concerned all our sense-perceptions are in the same boat, and must be treated on the same principle."[36]

From our perspective, the difficulty of integrating conscious experience and physical nature within one system is a challenge to be recognized and accepted, not swept under the rug by ignoring the primacy of direct experience and resorting to the surreptitious substitution or the fallacy of misplaced concreteness. One cost of the Blind Spot has been centuries of virtually no real progress on the mind-body problem. Although acknowledging the irreducibility of experience does not answer the difficulty, it opens the door to new approaches, ones that stay connected to the source of science in direct experience, as we discuss later.

What We've Learned So Far

We began this chapter by identifying a set of key elements of Blind Spot metaphysics. We can now summarize how they lead to the occlusion of experience through the following interlocking mistakes:

1. **Surreptitious substitution.** This is the replacement of concrete, tangible, and observable being with abstract and idealized mathematical constructs. Besides the parable of temperature, other examples we will discuss are substituting clock time for duration, nature at an instant for nature as process, computation for meaning, and information for consciousness. The surreptitious substitution is essentially the replacement of being with the products of a method for gaining a particular kind of knowledge. It's also the replacement of the life-world and nature outside the scientific workshop with what we manufacture inside the workshop.

2. **The fallacy of misplaced concreteness.** This is the error of mistaking the abstract for the concrete. It underlies the surreptitious substitution.

3. **Reification of structural invariants.** Science produces structural invariants through abstraction from experience in the scientific workshop. They include classification schemes, models, general propositions, logical systems, and mathematical laws and models. They comprise highly distilled residues of experience. Reification happens when they are regarded as essentially nonexperiential things or entities that constitute the objective fabric of reality.

4. **The amnesia of experience.** This happens when we become so caught up in surreptitious substitution, the fallacy of misplaced concreteness, and the reification of structural invariants that experience finally drops out of sight completely. It now resides in the Blind Spot we have created through misunderstanding the scientific method.

This list details in abstract terms our diagnosis of the underlying causes of the crisis of meaning in our global scientific worldview. We believe that these fundamental mistakes are also implicated in the existential crises that our scientific culture now faces.

Let's return to the parable of temperature for an illustration. If we say that how hot or cold something feels is subjective and apparent, whereas thermodynamic temperature (the average kinetic energy of atomic motion) is objective and real, we're thinking in terms of the bifurcation of nature. We're surreptitiously substituting an abstraction—a mathematical average, a single number taken as representative of a list of many numbers—for something concrete—an object of sense perception. We're committing the fallacy of misplaced concreteness by treating the abstraction as if it were concrete. We are thereby also reifying a structural invariant of experience. As a result, we've lost sight of direct experience as the source and sustenance of science. We've forgotten the entire process by which structural invariants such as thermodynamic temperature are extracted from but remain residues of direct experience in the scientific workshop. We've succumbed to the amnesia of experience. We are fully ensconced in the Blind Spot.

The irony is that as a result of human-caused global warming, the concrete reality of hot and cold has never been more tangible. At the same time, it is impossible to understand climate change and the Earth system without thermodynamics and statistical mechanics. We need to affirm the validity of both bodily experience and science without creating a division in our ideas about what is real and a hierarchy of knowledge based on that division. We feel the heat before we know how to measure it quantitatively. But we need both feeling and measurement to face the challenges of global warming.

If we refuse to allow nature to bifurcate in our thinking, our worldview changes. Hot and cold are real, no less than the average kinetic energy of particles. When hot and cold are found in nature, so too are particles having a wide range of kinetic energies, with most of the particles having energies near the middle of the range. Hot and cold are concrete; temperature is abstract. Better yet, hot and cold are relatively more concrete, and temperature is relatively more abstract, because to single out hot and cold from among the myriad deliverances of sensation and perception is already to abstract from the flux of events. That's why we say colloquially that the more you focus on being hot, the hotter you will feel. The beauty and power of science are that it enables us to intensify, prolong, and refine the abstraction already begun by sense perception, using the tools of logic, mathematics, and controlled experimentation in the scientific workshop. Not allowing nature to bifurcate in our thinking is the first step beyond the Blind Spot to a renewed vision of science.

Science doesn't lose its authority in this new vision; on the contrary, its authority gains a proper footing. Science is authoritative precisely because it's a highly refined form of experience. Its objectivity derives from our ability to abstract from direct experience and to make our abstractions objects of public knowledge. The powerful methods of science, visible in the parable of temperature, demonstrate how ascending cycles of abstraction, idealization, experimentation, and theorizing create stable public knowledge that communities of investigators can build on. This kind of knowledge enriches and enlarges human experience. Scientific knowledge is absolutely necessary to foster a vibrant and ecologically viable global civilization. But science must be properly understood so that nature isn't bifurcated and the life-world isn't degraded. Science must be properly embedded in a world outside the workshop focused on respect for nature instead of treating nature as something to subjugate. We need to keep this perspective in mind as we now dive deeper into the appearance of the Blind Spot in the history of science and science's modern incarnation.

2 The Ascending Spiral of Abstraction: Scientific Origins of the Blind Spot

It Made Sense at the Time

The Blind Spot emerged historically alongside science for a good reason. It was a natural response to the extraordinary success the new methods of science were delivering to European societies beginning in the seventeenth century. That was when communities of investigators across Europe started setting up different configurations of the scientific workshop in the form of laboratories, observatories, and learned societies that laid down norms for data collection, data analysis, modes of reasoning, and mathematical argumentation. As a result, in the space of just a few centuries, a remarkable series of advances in the understanding of key questions, such as the nature of motion, force, heat, light, and even the basics of biological development, was achieved. This knowledge was empirical and theoretical, allowing science to be put to use quickly in reshaping endeavors like agriculture, mining, trade, warfare, and even the conceptual and mathematical frameworks used for financing these activities. It was a time of unparalleled triumph for science. In the wake of such success, the constellation of elements making up the Blind Spot seemed to emerge naturally, and naturally be paired with, the root causes of that triumph.

This period between the late sixteenth and late nineteenth centuries is the era of classical physics. In this chapter, we focus on physics, because more than any other domain, its progress came to define the ideal form of what science could deliver. The triumph of classical physics from the seventeenth to the nineteenth centuries allowed the disembodied framework of the Blind Spot to form and take hold, and eventually to become dominant, so that it faded into the background as simply "what science says."

Thus, to understand how the Blind Spot limits scientific culture now, we must first understand how it came to seem natural for the culture then. We begin with the ancient Greek roots of science. It was during that remarkable period that key debates, and their essential polarities, were established. Because the current form of science emerged from European society (though it would owe much to global circulations of knowledge created by other societies) and because the recovery of Greek sources was the basis for the intellectual reawakening of the Scholastics and later the Renaissance, the conceptual frameworks of science trace back to the Greeks.[1] From there we will track the evolution of classical physics, and its all-important conceptions of motion and force, from the Scholastics in the thirteenth and fourteenth centuries on to Newton's unparalleled achievements. It was Newton who codified what a successful description of reality should look like in order to establish the existence of "laws of nature." We conclude with an exploration of the refinement and extensions of Newton's work by Joseph-Louis Lagrange and William Hamilton. The "analytical mechanics" they and others developed represented the crowning achievement of classical physics. It deployed a level of abstraction that seemed to offer convincing proof that the universe was indeed written in the language of mathematics.

The aim of this historical tour is to see how the philosophical elements of the Blind Spot—the bifurcation of nature, reductionism, objectivism, physicalism, the reification of mathematical entities, and a conception of experience as a derivative sideshow—both arose from and rode coattails with the achievements of classical physics. We also need to keep an eye out for how the amnesia of experience came as part and parcel of the Blind Spot and its assumed metaphysics for science. The chapters that follow explore how the Blind Spot metaphysics limits and distorts scientific thinking and practice today across a variety of fields, including physics. In this chapter, however, we focus on the scientific roots of the Blind Spot in order to discern how the surreptitious substitution, the fallacy of misplaced concreteness, the reification of structural invariants of experience, and the amnesia of experience took hold as classical physics became identified with the metaphysics of the Blind Spot.

Greek Ideas of Nature

"Every philosophy is tinged with the coloring of some imaginative background which never emerges explicitly into its train of reasoning."[2] When

Alfred North Whitehead wrote these words, he was speaking specifically of the Greeks and the role they played in establishing the roots of modern science. The imaginative background of the Blind Spot and its unspoken metaphysical conceptions can be traced to the Greeks and to monotheistic cultures of the Abrahamic traditions. It was the combination of the Greeks' belief in rational world models and the monotheistic vision of a rational God as the ultimate objective reference frame that would provide the soil out of which the mechanistic universe of classical physics would grow.[3] Thus, to trace the roots of the classical worldview of matter and motion, the one that would serve as a foundation for the Blind Spot, we begin with the pre-Socratic Greek philosophers and then advance across fifteen centuries of intellectual history to consider the medieval roots of science.[4]

In sketching the Greek legacy for both science and the Blind Spot metaphysics, we emphasize three key questions of Greek philosophy: (1) What is the world made out of? (2) How does it change (or not change if change is unreal)? (3) What role does mathematics play in the structure of the world? We give a few examples of how the Greeks addressed these questions in ways that would become crucial for the rise of physics.

The idea that the world is built from an original, unchanging substance defines the beginnings of Greek philosophy. The idea went hand in hand with thinking that natural things (in contrast to artificial ones) constitute a single "world of nature" and that what is common to all natural things is that they are made of a single underlying material.[5] These ideas were central to the Ionian thinkers of the sixth century BCE.

Thales of Miletus, the first of the Ionian philosophers, argued that water constituted the fundamental substance out of which everything is made. Anaximander, his fellow Milesian, however, regarded the primal stuff not as water, which is just one of many natural substances, but rather as something undifferentiated, indeterminate in quality, and spatially and temporally infinite, which he called "the Boundless." At the end of the sixth century BCE, Anaximenes combined aspects of these two theories. Like Anaximander, he regarded the primal stuff as extending infinitely in three dimensions, but he did not regard it as indeterminate in quality. Instead, he reverted to Thales's position that it is one specific natural substance but held that this substance is air or vapor. Not long after, Heraclitus would claim that fire is the substance underlying all things.

With their emphasis on a unique substance that constitutes all that is, these Greek philosophers opened the door to a line of inquiry that would continue down to our own time. Their claim was an ontological one, positing the existence of a foundational substratum underlying all aspects of nature.

About a century after Thales, Empedocles, who lived in Akragas, a Greek city in Sicily, composed a poem, "On Nature," that contains not just an ontology of nature but also a cosmogeny (a theory of the origin of the universe) and a theory of generation and destruction.[6] Instead of being made up of a single substance, everything is composed of four basic elements (Empedocles called them the "roots" of all things)—fire, earth, air, and water—and results from two forces, Love and Strife, which mix and separate the elements respectively.

Aristotle, in his *Metaphysics*, credits Empedocles for being the first philosopher to identify these four elements.[7] The most influential conception of them, however, comes from Aristotle himself.[8] He included them in a grand cosmological scheme that not only accounted for their combinations into the various material forms in the world but also explained their changes, including motion.

Aristotle argued for a geocentric universe in which the Earth is surrounded by nested crystalline spheres, each in motion and carrying a celestial object (sun, moon, planets, and stars) along with it. To account for the observed celestial motions, including the daily and annual rotation of the skies (of the Earth, really, daily around itself and annually around the sun), Aristotle proposed the existence of no fewer than fifty-six co-rotating nested spheres, imagining a truly complex system. It was not a mechanical system, however, but an organic one, because the world of nature for Aristotle is a world of self-moving things, as it was for the Ionian thinkers who preceded him.[9] The four elements—fire, earth, air, and water—belong to the sublunar realm, where generation, change, and decay are possible. The realm beyond the moon is perfect and changeless, with all celestial objects consisting of an incorruptible fifth element (called quintessence or "fifth essence"). The four terrestrial elements, however, are not the whole of nature by themselves. "Sensible qualities" (hot, cold, wet, dry) are added to the ultimate building blocks of the world. When combined in pairs, they underpin the behavior of the four elements. Water, for example, was wet and cold, while fire was hot and dry. What Aristotle was proposing was a cosmological model, an ontological listing of what existed in nature independent of human perception, as well as a description of how the items on that list interacted.

Every philosophical system needs a foil. For Aristotle's view of matter, that opposition came in the form of atomism. A century after Thales, Leucippus of Miletus and Democritus of Abdera argued that the universe is composed of an infinite number of minuscule and indivisible corpuscles of matter—atoms—moving randomly in a void. While the atoms themselves are unchanging, their motions and collisions give rise to the transient configurations that we perceive as the macroscopic world. These first atomists offered ingenious accounts of how the world of experience could be built from atomic motions. To us, what matters most is that their vision was reductionistic, materialistic, and mechanistic in the modern sense.[10] All the qualitative differences between things are attributed to differences in the shape, size, arrangement, and motion of the hypothetical atoms, and all of the substantial and qualitative changes that physical bodies are perceived to undergo are reduced to the motion and different combinations of atoms. In addition, the only way that the atoms can affect each other is through the effects they have when they come into contact with one another. There are no attractive or repulsive forces that can act at a distance. The universe is a lifeless mechanical system built from the motions of inert atoms that move according to rules that have nothing to do with life or mind or purposes. Chance and mechanical motion, not gods, rule the world.

More than a century later, atomism would be given a new face in the philosophy of Epicurus, for whom the universe was also an eternal infinite void populated by swarms of atoms in persistent motion. Because Epicurus was primarily interested in ethical questions related to how to live a good life, his atomism was meant to free people from their superstitious fears by showing them that the world and its features could be understood in terms of intrinsic properties that were without intention or agency. This meant that what we now call "secondary qualities," like taste and color, had no underlying existence. Only atoms and the void were ultimately real, with their properties of shape, size, and weight. As David Lindberg writes in his account of the Greek roots of science, for atomists there could be "no ruling mind, no divine providence, no destiny and no life after death."[11]

Epicurus did want to save human agency (free will), however, from a purely deterministic conception of the universe. To do this he introduced the "swerve" into his version of atomism.[12] As each atom moves, it shifts by the "least possible amount," adding an element of acausality to atomic motion. No force from an imposed cause drives the swerve. No prediction can be made of its direction. Although this addition of randomness would haunt Epicurus's version of atomistic philosophy, in its way it did anticipate the acausality that would come to underpin quantum phenomena.

Parallel to these ontological questions about the constituent elements of nature, the Greek conception of the role of mathematics in describing the world was equally important in the long road to classical physics. The triumph of abstraction in classical mechanics also had its roots in pre-Socratic Greece.

In 530 BCE, Pythagoras founded his school in Croton, Sicily, a community of mathematical mystics devoted to finding the numerical and geometrical blueprint underlying reality. Aristotle described their views (which he did not support) in this way: "[The Pythagoreans] took the elements of numbers to be the elements of all beings, and the whole of heaven to be harmony and number."[13]

R. G. Collingwood writes in *The Idea of Nature*, "Pythagoras broke new ground, with momentous consequences."[14] He proposed that the qualitative differences in nature are based on differences of geometrical structure or pattern rather than differences of stuff. Matter is that which can be shaped geometrically. To explain the nature of things, we must appeal to their mathematizable properties, specifically to their respective geometrical structures.

Acoustics and music theory provided the demonstration of this new viewpoint. The qualitative differences between one musical note and another do not depend on the material out of which the strings producing the notes are made but instead on the strings' rates of vibration. Pythagoras also discovered that musical notes with an agreeable or "harmonic" (a word the Pythagoreans invented) pitch to the ear have integer fractions of the length of the strings of a lyre. These relations, making up a musical scale constructed from fifths (3:2) and octaves (2:1), are an early instance of the discovery of how numerical relationships can be discerned in and through sense experience.

The Pythagorean breakthrough was to see form in things as making them be what they are. Form is essential, not matter—that which is capable of taking on form. In Collingwood's words, "Relatively to the behaviour of the things in which it exists, form is essence or nature. Relatively to the human mind that studies it, form is not perceptible, like the things that go to make up the natural world: it is intelligible."[15]

Plato (whom Aristotle regarded as a Pythagorean) elevated this conception of intelligible form into a principle of true being. A mathematical circle is perfectly circular, but a round plate is only approximately and imperfectly circular. Any bodily or material thing can only be approximate with regard to its perfect form. A material thing is also inherently transitory, always losing or falling away from its form, dying or becoming something else. But the intelligible form is fully and solely itself and never changes or dies. Its being is absolute. In addition, Plato conceived of form as transcendent. Its being does not belong to the perceptible world of nature but rather to an independent, intelligible world of pure forms.[16]

In Plato's cosmology, mathematical form is preeminent. Matter is that which is capable of receiving geometrical form, and the material world is an imperfect imitation of the intelligible world of forms. The elements (fire, earth, air, and water) are identified with mathematically distinct structures, specifically with four of the five regular geometrical solids (regular polygons meeting at the same three-dimensional angles and whose faces are all identical): the tetrahedron for fire, the cube for earth, the octahedron for air, and the icosahedron for water. The fifth so-called Platonic solid, the dodecahedron, is used for the universe as a whole because it most closely approximates the shape of a sphere. In addition, space is simply the receptacle of forms, and so it is no different from matter and does not correspond to anything in the intelligible world of forms. Time, however, is a copy of something in the intelligible world, namely, eternity. As Plato states in his dialogue *Timaeus*, time is a moving image of eternity.[17] "Eternity" does not mean infinite time that never ends. It means the absence of time or being outside of time. It means the total absence of change and passage. (Circular things come and go, but the mathematical circle does not reside in time and undergoes no change.) Time (change and passage) belongs to the natural world, not to the intelligible world of forms. Time comes into being with the creation of nature. Creation is the act of a divine craftsman or "demiurge," and nature is its handiwork. Hence, there is no time before creation, and creation is not an event in time. It is an eternal act.[18]

These Platonic and Pythagorean ideas—that matter is a mere receptacle for intelligible mathematical forms, that mathematics is the invisible skeleton of reality, that time is a moving and deceptive image of a realm in which there is no change or passage, that nature altogether is an imitation of a higher and perfect, intelligible order, and that nature is a system that

can be comprehended entirely from an outside, divine vantage point—
would have profound influences on the development of what we now call
theoretical physics. The capacity of filigreed mathematical abstractions to
be deployed in describing the world with the highest accuracy became the
bedrock of the triumph of classical physics. It would serve to push concrete perceptual experience to the background, a stance Plato had already
championed. As Lindberg writes, "Plato equated his forms with underlying
reality, while assigning derivative or secondary existence to the corporeal
world of sensible existence."[19]

It is important to note that the Greek philosophers were confronting the
particularities of their own historical moment and culture in their efforts to
move beyond dominant polytheistic narratives of the world and its order.
They were attempting to understand the world using reason, observation,
and experimentation. The techniques they developed, such as atomistic
reduction and the elevation of mathematics, would, however, outlast these
contingent urgencies and serve as one foundation for establishing the powerful shared language of inquiry we now call science. It is therefore with
good reason that we celebrate their innovations. The habits of thought
we associate with the Blind Spot, however, were also nascent in their philosophies, even if they were invoked for reasons particular to their own
age. These ideas would be called on to serve new purposes in the rise of
classical physics.

Toward Mechanics: The Medieval Search for Mathematical Motion

After the fall of the Western Roman Empire the tradition of "natural philosophy" established by the Greeks (though it was not called that at the
time) was taken up by the Arabic Islamic civilizations. There, vibrant discussions of issues raised by Aristotle continued. Advances in mathematics
and astronomy were also made that built on the work of Ptolemy and his
geocentric model of the universe. Although debate about the natural world
did continue to some degree in medieval Europe, it was not until Latin
translations of seminal texts in Arabic and Greek appeared in the twelfth
century that significant studies began again in earnest. Since we are mostly
interested in the development of classical mechanics from the seventeenth
to the nineteenth centuries, we first focus in what follows on the struggle
by late medieval scholars to formulate an understanding of motion, the

basis of all physics. Their work helped set the stage for the subsequent revolutionary advances by Galileo, Newton, and others.

The work done between the twelfth and the sixteenth centuries, however, must not be viewed solely under the lens of what came after. As Lindberg puts it, "We must never succumb to the temptation of supposing when we have identified the pieces of medieval physics appropriated by later ages, that we have figured out what medieval physicists themselves regarded as essential features of their discipline."[20] The scholastics were working under a view of the world that had not yet bifurcated into primary qualities of the external world "out there" versus secondary perceptual qualities, which would become so familiar to modern scientists. That future bifurcation of nature was what the revolutionary development of classical mechanics would help to bring about.

From Aristotle, the medieval scholars inherited the idea of "natural motion." In the realms including the moon and above, natural motion was the unceasing passage of bodies along perfectly circular paths. In the sublunar realm, however, motion could and would stop when an object reached its natural place (either downward toward the Earth, like a falling rock, or upward toward the sky, like fire). According to the Aristotelians, objects could be set into "violent motion" when a force acted on them (like a thrown rock), but the motion would stop once that force ceased to act.

For Aristotle and his followers, the kinds of motion that would come to dominate the concerns of physics after Galileo and Newton were just one of many kinds of change. It was the least important kind. The other three Aristotelian categories of change were generation and corruption, alteration, and augmentation and diminution. Each stood alongside "local motion" as a philosophical focus associated with the question of change that medieval scholastics endeavored to understand. Particularly interesting for our story is how only a few scholars saw the possibility of expressing local motion in terms of mathematics. Over the course of two centuries these scholars would struggle to translate intuitive notions of "quickness," "force," "resistance," and so on into mathematical quantities that could be manipulated and analyzed.

Gerard of Brussels attempted one of the first such analyses in his brief book *On Motion* (written between the late twelfth and mid-thirteenth century).[21] He recognized the appropriate split between kinematics and dynamics. Whereas dynamics is the study of what causes motion (what we now call forces), kinematics was a pure description of the motion itself. Bracketing (as Husserl might say) the kinematical relationships of distance, time, velocity, and acceleration from the conceptual apparatus needed to understand the causes of motion would prove to be fertile ground. Nevertheless, while Gerard saw there was a distinction to be made, he was not the one to develop the refinements needed to make progress in kinematics. That achievement was accomplished by a group of scholars at Oxford University's Merton College in the mid-1300s.

The Mertonians, a group that included Thomas Bradwardine, William of Heytesbury, Richard Swineshead, and others, developed the all-important technical vocabulary that allowed the key players in kinematics to be identified, conceptualized, and analyzed. This included distinguishing between uniform motion and nonuniform motion. The first was motion at a constant velocity. Velocity was itself a difficult concept, and the Mertonians' innovation was to recognize it as the distance covered in a set duration of time. They saw that nonuniform motion meant acceleration, where the velocity was changing with time. The Mertonians were even able to describe uniformly accelerated motion, where the velocity changed by equal amounts in equal units of time. By carefully articulating the meaning and relationship between each of these concepts, they created a language that dissected motion into its component parts and cases.

As the Mertonians' technical accomplishments reached universities across Europe, their ability to express change in quantitative terms was extended by other scholars. In France, Nicole Oresme developed a method of geometrical representation that would foreshadow the developments of Descartes and his analytic geometry two hundred years later.

Oresme found that he could represent the "intensity" of some property of an object, say heat, using a two-dimensional drawing similar to a modern graph. The "subject line" or "extension" could be drawn to represent the distance along the object (what we would call the x-axis on a modern graph). The "intensity" of the property at each point along the object's length was represented by a line perpendicular to the extension (what we would call the y-axis). Thus, Oresme could represent an iron rod that was cold on one end and hot on the other end through his proto-graph method by showing a horizontal line for the rod and a diagonal line rising from one tip marking the intensity of heat (what we would now call temperature).

After reading the Mertonians' kinematical work, Oresme found that he could represent motion in the same way. By making the subject line represent time and the intensity line represent distance, Oresme discovered a geometrical representation of motion that was ripe for the application of mathematical analysis. Using this representation, he and other scholars who followed were able to derive sophisticated kinematic results. Today these results are covered in the first lecture of a modern introductory physics course. In the medieval era, however, they represented an extremely advanced and hard-won understanding of the basic elements of motion. Classical physics would not have been possible without these developments.

Although the Mertonians and other scholars were also interested in dynamics—the causes of motions—they recognized that for this problem, the ground was muddied. The legacy of Aristotle's natural motion and his insistence that forces must act continuously on an object or it would grind to a halt posed a steep hurdle. The difficulty came in untangling exactly which mix of kinematical quantities was related to the imposition of a force. Medieval scholars stumbled toward different versions of an answer. In Paris, Jean Buridan coined the term *impetus* to describe a quality "whose nature it is to move the body in which it is impressed."[22] First, Buridan defined a quantitative expression for the impetus, with its strength depending on both the velocity (v) of a body and the amount of matter (m) it contained. Then he attempted to apply his impetus theory to projectile motion (a subject that vexed many medieval scholars) as well as planetary orbits. It is noteworthy that Buridan's impetus has the same definition as the quantity we now call momentum (mass multiplied with velocity, or mv), which would later reappear in Descartes' work and play a central role in Newton's dynamics. It's also noteworthy that Buridan conjectured what later became known as the law of inertia, usually attributed to Galileo: "This impetus would endure for an infinite time, if it were not diminished and corrupted by an opposed resistance or something tending to an opposed motion."[23] Although the role of Buridan's impetus was wholly different from that of momentum in Newton's account of force and motion, their identical mathematical form illustrates a transformation that was occurring in the late medieval period, echoing the work of the Pythagoreans two thousand years earlier: number and mathematizable quantity were becoming the primary language to represent the world and its changes.

Matter, Motion, and the Birth of Classical Physics

From the Greeks to the medieval scholastics, it took the West almost two millennia to arrive at a primitive quantitative description of motion. There were several reasons for the delay, but two stand out. First, Aristotle and his followers divided the cosmos into two realms, each satisfying very different rules. While celestial objects moved in perpetual circular motions for all eternity, Aristotle's division of terrestrial motion into "natural" and "forced" motion obscured the notion of inertia. Second, although the Mertonians and a few other medieval scholars took the first steps toward identifying the anatomy of motion through their kinematical studies, a force-based description of the causes of motion still eluded them.

All this would begin to change in the 1600s with Galileo, Descartes, and Newton. From there, progress increased at a rapid pace through the work of Leonhard Euler, Joseph-Louis Lagrange, William Rowan Hamilton, and others, reaching a stunning level of sophistication and mathematical abstraction by the end of the 1800s with the work of Henri Poincaré. Classical mechanics unified the physics of the celestial and terrestrial realms into a single coherent body of knowledge, firmly rooted in mathematics and with solid empirical validation.

In *Science and the Modern World*, Whitehead reminds us that what was most radical about the era that gave birth to classical mechanics was not a straightforward embrace of reason but rather an exceptional recoil from reason and an appeal to brute fact.[24] This claim may seem strange, given the enormous sophistication of the mathematical inventions that made classical physics possible. Whitehead's point, however, is that the makers of classical physics, particularly Galileo, were abandoning a metaphysics that envisioned the world as an ordered whole, which the mind of the Creator infused with purpose. These natural philosophers stuck to the "irreducible and stubborn facts" instead of bending their inventions to the rational requirements of the Aristotelian and medieval theological order. They were intent on developing an account of matter and motion that worked, explained how things happen, not why, and matched observations and experiments. An overarching philosophical framework that explained why things happen was not required. In the absence of such a framework, the metaphysics of the Blind Spot—an assumed objective ontology of mathematizable matter moving blindly in the void—eventually slipped into place as the de facto metaphysics of nature and philosophy of science.

As we will see, the Blind Spot's manifestation in classical physics was built on Greek themes. In particular, the essential role of mathematics as an invisible scaffolding behind the physical world's form and change was central to the progress of classical physics. The increased sophistication and applicability of mathematical abstraction in classical physics seemed to echo Plato's doctrine of intelligible forms. From Newton's geometrical calculus to the hyperdimensional phase space of Hamiltonian dynamics, classical physics is a story of mathematical constructions that successfully built on, surpassed, and then forgot the role of lived experience. Classical physics is also the story of surreptitious substitution, as newly formulated mathematical laws were thought to be more fundamental than the lived world they were believed to rule over.

Galileo set the tone for the new age by demanding that all "natural philosophy" pay direct attention to nature through the specific methods of experimentation. Employing a new instrument, the telescope, and using carefully crafted experimental setups of blocks sliding down inclined planes or the careful examination of pendular motion, Galileo developed a new standard for both questioning nature and extracting its answers. Through his experimental studies, some started as a young man and others carried out during the last years of his life under house arrest, Galileo set mechanics on a new course. In particular, he was able to see what all others had missed, though he may have been aware of Buridan's discoveries, including his law of inertia. Uniform motion—a body traveling at constant velocity—was motion that required no force. Left unhindered, an object moving at constant velocity would continue moving at that velocity forever. It was only the application of a force that interrupted uniform motion and brought change. Galileo had (re)discovered the law of inertia, giving it a proper kinematic context.

Galileo's experimental work exemplified the creation of the scientific workshop.[25] His insights did not come primarily from watching horses drawing carts in the marketplace or hammers rebounding from a smith's anvil. Although those commonplace settings could have been the source of insights, they were too "noisy" as a setting for extracting structural invariants from experience. Instead, Galileo's insights occurred in a new, specific, and artificial kind of location. He created a physics laboratory where nature could be "put under constraint and vexed" (in Francis Bacon's problematic phrase).[26] Through the use of wooden blocks and inclined planes, Galileo isolated the experiences he wanted to attend to. In doing so, he forged the first links in a new, quantitative study of dynamics. Like our parable of the invention of temperature that ultimately led to thermodynamics and statistical physics, Galileo's experimental work in kinematics marks the initial steps that led to classical physics.

Newton would take the next epoch-making step with his three laws of mechanics. But between Galileo and Newton, René Descartes provided an essential mathematical key through the development of analytic geometry. Descartes created a formalism for representing the space through which motion occurs. In his schema, space was separated into three orthogonal directions, which could be thought of as breadth, width, and height. Each direction was represented as a line—an axis—delineated with coordinate markings: x for breadth, y for width, and z for height. Thus, every point in space could be labeled with a triplet of numbers (x,y,z) that gave the point's coordinates.

It was a potent abstraction. Using this address system for space, Descartes provided the means for representing geometric constructions (lines, curves, surfaces, solids) in terms of functions of the coordinates: $f(x,y,z)$. All future progress in classical mechanics would rest on this abstract method for representing space and the changes occurring within it. By so closely linking the experienced space of everyday life with the abstracted space of this address system, Descartes' method established a foundational image for the reification of mathematics. Here was a mathematical formalism that lifted the rooms and roads, boxes and balls of daily life into a higher, perfect realization in pure abstraction. But it was also abstraction with tangible power, enabling workshop constructions (a theory of force and motion) to be put to work in the world outside the workshop in the form of powerful new technologies (such as more accurate canons).

Descartes also made an important contribution to mechanics in his recognition of the importance of momentum or "amount of motion." He saw that momentum should be mathematically described as mass times velocity (mv). In doing so, he identified the correct quantity to focus on in the description of dynamics. For example, in the interaction between bodies, such as a collision between two billiard balls, the total momentum of the moving spheres (the momentum of sphere 1 plus that of sphere 2) before and after the collision must be the same. The total momentum is unchanging. It is conserved in the interaction. Such conservation laws would come to play a critical role in the progress of classical physics (and indeed in all of theoretical physics).

Newton's contributions to science were vast. They ranged from seminal studies in optics and the law of universal gravity to the independent coinvention of calculus (the other inventor being Gottfried Wilhelm Leibniz). What matters most for our story was Newton's explicit definition of the effect of forces. Building on Galileo's principle of inertia and Descartes' concept of momentum, Newton finally understood the explicit role of forces as influences that change states of motion. A force changes an object's motion by changing its momentum. (In symbols, $F = d\,(mv)/dt$), which is Leibniz's notation, d, is the derivative, a measure of change of a quantity. Thus, this formula states that the force F changes the object's momentum in time.) If the mass of the object is constant, then the force changes only the object's velocity, producing an acceleration, and giving us the famous formula, $F = ma = m\,(dv/dt)$. This "force law" is universal, meaning that it describes the effect of any force exerted on any object in any location, cosmic or terrestrial. Finally, Newton also saw that forces must always come in pairs between the body exerting the force and the body "reacting," that is, responding to it.

The law of inertia, the force law (law of mass, velocity and acceleration), and the law of reaction forces (third law of motion): together these principles set out Newton's three laws of motion. Their success at describing the world made them the blueprint for what the new age of science could achieve. Newton's mechanics allowed scientists to describe the path of planets and comets, build stronger bridges, and make more accurate weapons.

Newton called his principles both "axioms" and "laws." An axiom is a foundational and established principle. Newton's use of the term alludes to a body of earlier work on principles governing motion under impact (impressed force). The term *laws of motion* had been in use for these principles since the late 1600s. Descartes had earlier used the term *laws of nature* for the principles of motion. He and Newton did not consider these principles to be mere statements about regularities. Rather, they govern the world. The natural world "obeys" the laws of nature. God imposes natural laws on the physical world (and moral laws on human beings). God is the primary cause; his laws are the secondary cause.[27] When Newton used the term *laws*, he implied that his laws of motion are the real governing structure of God's creation.

This early connection between classical mechanics and theism should not be understated. Even after science abandoned the idea of God (especially a deist God), the idea would still unconsciously shape the Blind Spot through its emphasis on objectivist ontologies—the world grasped objectively from God's outside perspective. As Albert Einstein said centuries later, "I want to know how God created this world. . . . I want to know His thoughts, the rest are details."[28]

In Newton's case, the connection between classical mechanics and theism becomes apparent in his conceptual steps to describe motion mathematically. He needed to create a new conceptual framework for the stage on which natural phenomena unfold. Since motion links space with time, Newton needed to define both. He thus defined "absolute time" as flowing uniformly throughout the universe, oblivious to anything or anyone. As we will see in the next chapter, Newtonian time was a steady marker for change, amenable to a rigid mathematical description. A uniform "absolute space" was just as essential for his equations (his laws) to make sense. In the *Scholium Generale* (General Scholium), appended to his *Philosophiae Naturalis Principia Mathematica* (known as *Principia*), Newton presents his arguments for absolute space as a pure void, a steady marker for distances between points. Like absolute time, it was also amenable to rigid mathematical description. Much of the *Scholium* rests on dynamical arguments. Underlying these justifications, however, we find that Newton's ontological framework equates absolute space and time—the stage of dynamics—with attributes of God.[29] In his earlier work, *Opticks*, Newton had already stated that absolute space is the "sensory" or "sensorium" of God, so that the divine mind perceives all bodies in space at once: "There is a being incorporeal, living, intelligent, omnipresent, who in infinite space, as it were his sensory, sees the things themselves intimately, and thoroughly perceives them, and comprehends them wholly by their immediate presence to himself."[30] Although later ages would drive God out of this description of space, the Newtonian God's absolute perspective, its "view from nowhere" or rather "view from everywhere," its "absolute conception of the world," would become lodged as an unquestioned assumption in the minds of physicists.[31] It would form the ideal of a perfectly objective vantage point for viewing the world as-it-is in-itself, a vantage point only science can provide. It would become part of the Blind Spot.

From Forces to Phase Space: The Triumph of Abstraction in Classical Physics

For all their power, Newton's laws were cumbersome to work with. Consider a problem familiar to any physics student. A small block slides down a frictionless wedge, which itself can move along the surface of a revolving turntable. Solving this problem requires knowing the initial position of the block in three-dimensional space (labeled as x_0, y_0, z_0), and then determining all the forces acting on it in each coordinate direction. Finally, differential equations determined from Newton's laws must be integrated (solved) to get the motion in each direction as a function of time: that is, $x(t)$, $y(t)$, $z(t)$. To properly solve the problem, the forces on the wedge and the block must each be resolved in each direction, including the "noninertial" forces from the rotating turntable. As most physics students can attest, correctly parsing the forces and integrating the resultant equations requires considerable experience and skill. In addition, if there are further constraints imposed on the bodies—for example, if a series of blocks is joined by a network of flexible rods—then the force descriptions need to include these constraints as well. Going further, when considering a large collection of objects like a box full of atoms (a gas), the Newtonian approach requires solving equations for every atom. One must start with all atomic initial positions and velocities and with all forces between them resolved in three directions. Calculating the future state for even a tiny box of gas would require solving trillions and trillions of equations. Thus, while Newton's program for the determination of motion in principle could be carried out for any collection of objects, in practice many situations are simply too complex to solve or even gain insight into. In addition, as we will see in the next chapter, this inability to directly solve Newton's equations for large, complex systems has profound implications for our understanding of time.

As succeeding generations of physicists considered the science of mechanics, they began to search for other, deeper principles at work shaping mass and motion. These principles would be found in a series of amazing advances that vastly enlarged the reach of mechanics while simultaneously casting it into ever higher levels of mathematical abstraction.

The most important advance in this era rested in the idea that nature, in some sense, is economical. As systems go through their changes, they attempt to keep some aspect of their motion to a minimum. Such "minimum" concepts had been expressed before in more restricted fashion a number of times. In 60 CE, for example, Heron of Alexandria saw that light always "chooses" the shortest path when it is reflected off a surface. Almost two millennia later, Pierre Fermat reconsidered Heron's result but saw that it is really a matter of light taking the least time in its travel from source to mirror to detector. In addition, in the years around Newton, many physicists and mathematicians worked to solve the famous "brachistochrone" problem, trying to find the proper shape of a hanging wire that minimized the travel time for a bead sliding down along its length. These and other examples served as hints that nature seems to choose "extremal" paths, meaning that it makes some property of motion either a minimum or a maximum. In turn this implies that some universal organizing principle is at work related to what one could call an economy of motion.

The first expression of a new general extremum principle for mechanics came with Pierre Louis Maupertuis, the child of a wealthy privateer. In the 1740s, Maupertuis discovered the "principle of least action." The "action" was a new quantity he introduced into mechanics. It was the product of an object's mass (m), velocity (v), and the distance along its path of motion (d). Maupertuis suggested that the action could be written as $m \times v \times d$, and he expressed his new principle in the simplest terms: "Whenever any change takes place in Nature, the amount of action expended in this change is always the smallest possible."[32] By finding the right physical quantity, the action, and the right idea, its minimization, Maupertuis pointed physics in a direction beyond Newton's more cumbersome version of mechanics. But he was not adept enough to translate the statement of a new general principle into a mathematically tractable formalism with general applicability. That would require a remarkable group of mathematical physicists, including Jean D'Alembert, Leonhard Euler, and Euler's successor as director of mathematics at the Prussian Academy of Sciences in Berlin, Joseph-Louis Lagrange.

Lagrange and the others found a mathematically elegant way to express the action of a system of objects across the entire path of its evolution. In this new program for mechanics, the physicist specifies the initial coordinates (x_0, y_0, z_0) and final coordinates (x_f, y_f, z_f) for the path (where the motion starts and ends). The total action is then summed up along the path. In other words, the total action is the integral of mass times velocity across the distance traveled. Finally, by using the "calculus of variations," the actions for all possible paths between the two fixed endpoints could be simultaneously examined. Remarkably (and beautifully), the path that nature chooses will be the one for which the action is "stationary," that is, the path for which the action is either a minimum or a maximum value. In practice, the extremum tends to be a minimum in mechanics. Thus, nature is said to express a principle of economy in its organization.

To carry out the procedure described, Lagrange created a new mathematical object that would shape the entire course of physics, both classical and quantum, touching every applicable domain from electromagnetic fields to Einstein's general theory of gravity. This object, now called the *Lagrangian*, was a single function (a "scalar") and thus much easier to use and formulate than Newton's three equations for three directions of motion, each formulated for every object in a system. Equally important, the Lagrangian was expressed in terms of the energies in a system.

Energy was a new idea for physicists in the late eighteenth century. Like momentum and force before it, the specific and appropriate mathematical definition of energy took time for physicists to pin down. Part of the difficulty was due to the energy in a system taking different forms. First, there is the energy locked up in the system's motion. This was called *kinetic energy*. ("Kinetic" comes from *kínsi*, the Greek word for motion, as in "cinema.") For a single particle, it was expressed as mass times velocity squared (or $\frac{1}{2} \times m \times v^2$). There is also the energy locked in the spatial arrangement or configuration of the objects comprising the system. This *potential energy* depends on the nature of the interactions (or forces) between the objects (such as through gravitational or electric fields). The potential energy was expressed as a function $U(x,y,z)$ that had to be specified and could be different for different kinds of interaction. For example, an object suspended from the ground at a certain height H has a gravitational potential energy that depends linearly on H. If let go, the object will fall to the ground, a transformation that can be described as the conversion of potential energy into kinetic energy.

The fact that the Lagrangian, through its expression in terms of a combination of different forms of energy, is a system-wide description is also important. The Lagrangian embraces collections of bodies as a whole. It can represent them as a single mathematical object. The extremal value of the action, which was also a system-wide description, comes from minimizing the value of the Lagrangian as the system follows a path from initial to final position.

For the Lagrangian to work, a very different idea of what coordinates represent was required. Recall that in Newtonian physics, a material body has a location in physical space described by the Cartesian coordinates x, y, and z. The forces on the body are written down in each direction, allowing Newton's equations of motion in these directions to be solved. Because Lagrange's program is a systems-level approach, it "generalizes" the meaning of coordinates. Lagrange's generalized coordinates are specific to the system and can express whatever constraints define it. As an example, consider a compound pendulum consisting of three weights, each connected by a rigid rod with hinges at the position of the weights. Each weight will swing back and forth like a simple pendulum, but the hinge points themselves are also swinging. For this problem, Lagrange's generalized coordinates consist of the angles each rod makes with the vertical. The Lagrangian for this problem also expresses the kinetic and potential energies of the system in terms of these angles, making this formulation far easier to analyze. What matters for our story is that Lagrange's generalized coordinates reimagine space in terms of the system itself rather than a rigid frame of reference pinned to the universe (or to God). In this way, they represent an abstraction of the meaning of coordinates and of space. Thus, the Lagrangian program takes a step away from the Cartesian framework, which, of course, was already an abstraction of the bodily experience of space.

We should note that the Lagrangian formalism always leads to a set of dynamical equations that recovers Newton's laws, as it must since those laws always hold. Nevertheless, Lagrange's approach, with its elegance and simplicity, allowed classical mechanics to reach far beyond Newton, greatly expanding its applicability and insight.

The next advance in classical mechanics came in 1833, when Irishman William Rowan Hamilton reformulated the principle of least action by further abstracting the representation of coordinates and space. He did this by making momentum itself a new kind of coordinate, on equal footing with position.

In Newtonian mechanics, forces change a body's momentum. When an object has constant mass, for example, this means that forces act to change the object's velocity, producing an acceleration a, expressed mathematically as $a = dv/dt$ (recall that acceleration is a measure of how velocity changes in time). Velocity, however, is nothing more than a measure of how position changes in time, $v = dx/dt$. This means that momentum is not really a quantity independent of position.

In Hamilton's formulation of mechanics, however, the momentum of an object is treated as an independent coordinate just like position. For just a single particle, this meant that the space the system "lived in" was no longer that of the usual three Cartesian dimensions. Instead, Hamilton's version of dynamics takes place in six dimensions, defining what would later be called "phase space." There are three dimensions for position (x, y, and z) and three dimensions for momentum (mvx, mvy, and mvz). For systems more complex than a single particle, both the momentum and position (configuration) coordinates can be further generalized to capture constraints within the system just as in the Lagrangian framework. The dynamical representation of the system, however, always exists in a "hyperspace" of more than three dimensions.

Over time, the tools developed for analyzing Hamilton's version of mechanics grew in mathematical sophistication, particularly through the work of Henri Poincaré in the late nineteenth and early twentieth centuries. As he and others understood, even when the final dynamical equations cannot be solved due to their complexity, Hamilton's recasting of mass and motion in terms of phase space still allows for powerful insights to be gained into the behavior of the system under study. For example, as the system evolves in time, it traces out a trajectory in its own multidimensional phase space of generalized momentum and generalized coordinates. The shape that trajectory fills out over time holds important clues to dynamics, for example, by revealing which quantities are unchanging (that is, conserved) in time. In this way, the mathematics of these hyperdimensional objects can be directly translated into insights about the real-world motion of planets or particles or projectiles.

The fruits of the Lagrangian and Hamiltonian formulations of classical mechanics were impressive in both their range and power. When Michael Faraday and James Clerk Maxwell in the middle of the nineteenth century introduced electromagnetic fields extending throughout space into the language of physics, these entities posed a challenge for the Newtonian view of a universe built of only particles and forces. A field is a physical quantity that can vary continuously in space and time. Faraday realized their importance during his studies of magnetism; he determined that electric and magnetic fields determine the motion of particles and carry energy, so they must have their own physical reality. Maxwell then provided a unified field theory with equations that form the foundation of classical electromagnetism. Eventually it was found that these fields could be elegantly embraced in the Lagrangian framework, as could continuous media such as elastic surfaces. In this way, phenomena related to light, which Maxwell had found to be traveling electromagnetic waves, were also embraced within the full analytical tool kit of classical physics. Furthermore, as we will see in the following chapters, by the end of the nineteenth century, atoms made a triumphant reappearance in physics through the work of Ludwig Boltzmann and Josiah Gibbs, who made wide use of phase space in their statistical mechanics. Here again, the Lagrangian and Hamiltonian formalisms were decisive in allowing physics to embrace a systems view of trillions upon trillions of interacting atoms.

The Triumph of Classical Physics

As the twentieth century began, the gossamer hyperdimensional abstractions of classical analytical mechanics appeared to embrace all of nature from the smallest atoms to invisible fields pervading space. The vast power and reach of the extraordinary endeavor of physics served as a spectacular vindication of human reason. Classical physics served as a model for the triumph of science altogether.

We have sketched the history of classical physics to bring to the fore how physics came to be associated with, and understood in terms of, the blind spot worldview, as outlined in the previous chapter. To conclude this chapter, we single out the association of classical physics with the following elements of the Blind Spot: the bifurcation of nature, reductionism, objectivism, and the reification of mathematical entities. We consider each of them in turn.

The bifurcation of nature. This is the idea that microphysical entities (atoms, light waves) exist externally and objectively, whereas perceptual qualities like color, or hot and cold, are subjective appearances and exist in the mind. Classical physics, of course, was the main impetus for this worldview, and the early modern physicists (Galileo, Descartes, Newton) were its main architects. Nevertheless, as Whitehead argued, the metaphysics of the bifurcation of nature does not logically follow from classical physics as an explanatory framework.[33] Instead, that metaphysics gets superimposed onto physics. As a result, the mind is banished from nature. As we will see in the chapters to come, however, the mind returns to haunt physics as it grapples with the problems of time, matter, and cosmology in the twentieth and twenty-first centuries.

Reductionism. The extensive use of the method of reduction—reducing phenomena to their component parts—was so successful in classical physics that it became elevated in the minds of many from a method to a fundamental principle of being. According to that principle, which we can call "smallism," small things and their properties are more fundamental than the large things they constitute. This goes hand in hand with a principle of knowledge called "microreduction": the best way to understand a system is to break it up into its elements and explain the properties of the whole in terms of the properties of its parts. In short, reduction, a useful and indeed powerful method for certain purposes in certain contexts, was turned into a universal "ism"—reductionism. This "ism" has had, and continues to have, broad philosophical and ethical consequences, as we will see throughout the book, particularly when we discuss the planetary climate crisis in the final chapter.

Objectivism. Newton's need to imagine a fundamental frame of reference for distinguishing between inertial and noninertial motion led him to introduce notions of absolute space and time. These had theological connotations for him. They also served to establish an objectivist outlook, according to which science is supposed to provide a God's-eye view of the universe. "Objective" does not simply mean independent of individual subjectivity, publicly checkable, and invariant under transformations of perspectives. It means how things are from the God's-eye view from nowhere and nowhen.

Reification of mathematical entities. The remarkable power of the Lagrangian and Hamiltonian formulations helped fuel the reification of mathematical entities as a key element of Blind Spot metaphysics. There is an enchanting economy and beauty to these formulations of celestial mechanics or particle dynamics that seem almost supernatural in their efficacy.[34] This efficacy was taken to support, or was understood in terms of, the Platonic conception of mathematical entities as what are truly real. Mathematics was taken to be the ideal skeleton of the world, instead of being a human creation born of the workshop and our cognitive capacity to abstract from lived experience.

Given this powerful worldview, it is easy to see how classical physics became the archetype of a scientific understanding of the world: science should ideally be perfectly objective and universal, untainted by human experience. This way of thinking led to rapid progress in considering the class of problems that classical physics was suited to address. But that broad class of problems left out large parts of nature and human experience. As science climbed to ever more rarified peaks of mathematical abstraction, the surreptitious substitution took hold, so that the constant foothold of direct experience required for the climb would be forgotten. This forgetfulness is the amnesia of experience we wrote about in chapter 1.

To begin our recovery from that amnesia, it is important to see how, across the whole history of classical physics, its success has always remained grounded in lived experience. While the surreptitious substitution was unconsciously leading scientists and philosophers to forget the ground of their inquiry, they always had to rely on that ground of direct experience.

For example, the medieval scholars and their conceptual struggles in quantifying the kinematics of motion illustrate the difficult path required to extract structural invariants from embodied experience. Human beings have experiences of motion by living in the world through their bodies. In the era just prior to classical physics, these experiences served as the intuitive background to find, for example, a proper mathematical definition of velocity from less explicit notions of "quickness" or "rapidity" or "fastness."

The subsequent progress in classical physics would echo these struggles, as natural philosophers proposed and explored formal, mathematically tractable definitions for inertia, momentum, and force. After Newton, a new spiral of abstraction would begin as succeeding generations of physicists would define new mathematical quantities like energy and action whose meanings lay further from embodied experience and yet provided even greater predictive power. So powerful were such abstractions that, in time, the fact that they remained residues of experience was lost sight of and forgotten. As we've followed this ascending spiral in this chapter, we can see how the mathematical abstractions of classical physics, at their height, came to provide a justification for a commonsense acceptance of the Blind Spot's metaphysical prerogatives.

As happened with the bifurcation of nature, reductionism, objectivism, and the reification of mathematics, so too would the other elements of the Blind Spot metaphysics (physicalism and epiphenomenalism about experience) come to be grafted onto assumptions about what science is and what it says. It all seemed to make good sense, though there have always been important critics and protesters along the way, including within science itself. Nevertheless, many scientists, particularly physicists, came to dream that all of nature was within their grasp. But nature, "as that which we experience," to modify Whitehead's phrase, had other ideas, as we will see.

II Cosmos

3 Time

Human Time

The alarm rings. Reluctantly, you slowly shake off the residual grip of slumber and start getting ready to greet the day. You gear up, allowing your senses to bring in the world. To awaken is to leave behind the imagined time of dreams and return to the inexorable flow of waking time.

That time takes you along as your day unfolds. An internal narrative develops with you as the protagonist, centered on your first-person perspective, your subjective sense of being in the world. You look at the mirror and see yourself, the same as yesterday but not the same. Every day is a little different. You, the world, and your experience are always present yet always changing as your future emerges and your past recedes.

To be human is to contend with the experience of passage and change. This experience lies at the heart of what we call time. We experience both the present moment (our sense of now) and its changing (its inexorable passing). We experience passage and change as irreversible: we grow older, not younger, and we can't restore past moments and make them now again. Contrary to space, where we can move backward and forward at will, we can't control the flow of time. We age, and the environment around us changes, either naturally or as the result of our actions. What we can control, to a certain extent, is what we do with our existence as it unfolds in time.

To be human is not just to experience irreversible passage and change but also to remember the past and anticipate the future and to parse time according to them in thought and action, word and deed. We order time sequentially to organize our lives: our birth, childhood, adulthood, old age, death, and the memories of others who have departed. Calendars are essential to us. We mark days and events of significance in our lives, often celebrating their cyclic repetition: birthdays, graduations, wedding anniversaries. Calendars based on significant astronomical events, such as the phases of the moon or the orbital period of the Earth around the sun—the solar year—are tools designed to give us a quantitative sense of the passage of time, defining what's in the past and how far back it is, and what's in the future and how far ahead it is.

Thus, human time is Janus-faced: one face looks inward, to the feeling of irrevocable passage, and the other face looks outward, to what Aristotle called time as a measure of change and motion.[1] Human time has probably always been this way, ever since we first became conscious of the inevitability of our own death and noticed the phases of the moon and the regular movement of the sun and the stars.

Lived Time and Clock Time

The Blind Spot appears in the opposition between these two ways of thinking about time: as lived time (experiential time) and as clock time (time as what a clock measures). The first thinker to understand this was the French philosopher Henri Bergson.

Bergson's first book, *Time and Free Will: An Essay on the Immediate Data of Consciousness*, written as his doctoral thesis, was published in 1889 when he was thirty years old.[2] Its key idea is that time is not space. When we think of time as a series of mutually external points, we spatialize it. More precisely, we use properties of space that can be expressed through mathematical language, such as geometrical points and lines, to mentally represent time. We conceptualize time as a succession of discrete, homogeneous, and identical units (like seconds). This is clock time. But we never experience time this way. An hour in the dentist's chair is very different from an hour over a glass of wine with friends. A group of runners may finish a 13.1-mile half-marathon race in two hours, but the passage of those two hours will vary greatly to each runner. This is lived time. For Bergson, lived time is real time, and clock time is an abstraction. Lived time is becoming. It is continuous, irreversible, and asymmetrical (the child becomes an adult, not the other way around). Spatialized time—clock time—lacks becoming. There is no becoming in a discrete succession of homogeneous and identical units (points on a line, seconds on a clock). Once we spatialize time—or, more precisely, geometrize it—we lose becoming. We lose the qualities of continuity (time as made up of overlapping and changing phases rather than discrete units), heterogeneity (each phase is qualitatively unique), and interpenetration (past, present, and future phases bleed into each other). In a word, we lose what Bergson calls "duration."

Music and dance are good examples for understanding duration. Melodies and dances exist only in duration. They have no being either at an instant or as a series of discrete moments. Each note in a melody has its own distinct individuality while blending with the other notes and silences that come before and after. Every motion and movement in a dance stands out while merging with the other ones. Past notes and movements linger in the present ones, and future notes and movements already seep into the ones happening now. Even melodies and dances that emulate discrete seriality cannot escape having their distinct elements meld into the intervening silences and pauses, and thereby also into one another. Melodies and dances are fundamentally durational.

Bergson is not opposed to clocks and measurement. He is opposed to the surreptitious substitution of clock time for duration, of spatial quantities for temporal qualities. He opposes the idea that time as measured by a clock is objectively real, whereas duration is merely psychological. On the contrary, as Whitehead also thought, nature as passage, as sheer becoming, is given in duration, and duration is the wellspring for constructing time systems using clocks.[3] Thinking that clock time is physically real, but duration exists only in the mind, is an example of the bifurcation of nature, of dividing nature into external physical reality versus internal subjective appearance. Indeed, as Isabelle Stengers notes, the division between physical nature at an instant, considered as objectively real, and the lived experience of passage and duration, considered as merely subjective, provides "the most complete example of the bifurcation of nature."[4] This way of thinking also exhibits the fallacy of misplaced concreteness, of mistaking the abstract (clock time) for the concretely real (becoming). For Bergson and Whitehead, what is concretely real is the passage of nature given in duration, and clock time is an abstraction from the passage of events.

The Blind Spot arrives when we objectify clock time and regard it as the only real time, while forgetting its necessary source in the concrete reality of passage. Time as passage is given to us in the experience of duration.

This is what Bergson saw so clearly. We need to appreciate the force of his argument that measurement presupposes duration while duration eludes measurement. We turn to that argument now.

Bergson on Duration

Bergson insists that "duration properly so-called cannot be measured" and that "as soon as we try to measure it, we unwittingly replace it by space."[5] The key to understanding what he means is to see how measurement works.

To measure something, we need to stipulate the unit of measurement in terms of a standard. For example, the standard meter was once stipulated to be the length of a particular platinum bar kept in Paris. Now it is defined as the length of the path light travels in a vacuum over an extremely short time interval (1/299,792,458 of a second) as measured by an atomic clock. Notice that the standard meter, a measure of length, itself has a length (the length of the platinum bar, the length of the path of light). We use length to measure length and volume to measure volume. Thus, the standard unit itself exemplifies the property it measures.

Let's apply this to time. As Aristotle saw, we use time to measure time, but then we translate time to space. Suppose we want to measure the time it takes for a body to move from one place to another. To measure this motion, the ancient Greeks (and many other cultures before them) realized that we can use a parallel motion, such as the motion of a shadow on a sundial (or the flow of water in a water clock if the sun isn't out). As the body moves, so too does the shadow, and we can perceive and register the correlation of these two motions and changes of place. If we number the positions of the shadow on the sundial, we can order them in terms of before and after, despite their being all simultaneously present. Thus, we say that position 5 is before position 6 and after position 4. In doing so, we translate from time—from before and after for the parallel motions—to numbered relative positions existing all at once in space. As Aristotle states in his *Physics*, "But whenever there is a before and an after, then we say there is time, for this is time: a number of motion fitting along the before-and-after."[6] The before and after are given by the motion of the shadow on the sundial, and the number of motion is given by the relative positions. The key thing to notice is that to measure time, we must use time, but in constructing a time standard—clock time—we spatialize time.

What Bergson wants us to see is that this procedure will not work for duration. If duration is to be measurable by a clock, the clock itself must have duration. It must be an enduring temporal entity. It must exemplify the property it is supposed to measure. Of course, we think of clocks as enduring temporal things. But Bergson asks us to take a closer look. Any state of the clock—any position of the oscillating pendulum in his example—is external to every other one, like the points on a line or the numbers on a clock face.[7] A clock can be described as a finite-state machine in which every state is external to every other one. Every state is a position in space juxtaposed to another position in space. Every state is simply now, with nothing left of the past. The past states do not endure in the present state. The preceding oscillation of the pendulum or chime of the bell is not held together with the present one while being apprehended as past in relation to it. *We* hold them together in our memory, but the *clock* itself cannot do this. Without such a holding together of the past and the present, however, duration can't be registered. All that can be recorded is one nonoverlapping state and then another one, but the sequence as such can't be registered because that requires memory. Memory, however, in the sense of the immediate retention of the past as it slips away, is part of duration. Every duration contains within its "now" the trailing threads of the immediate past. Husserl called this kind of memory "primary memory" or "retention," and he said that the apprehension of "now" is the head attached to the comet's tail of retentions.[8] The clock, however, has no memory. It lacks duration and therefore cannot measure it.

Bergson is not denying that we can measure time. On the contrary, his point is that *clocks don't measure time; we do.* A clock face showing 10:59 and then showing 11:00 is not measuring time. Measurement requires that we look at the clock, read the numbers, and notice there is a change. We must hold the prior states of the clock in our memory, in our durational consciousness. Take away the measurer's memory and you no longer have a measurement of time. Clock time presupposes lived time.

Bergson recognizes that we measure time with increasing precision. But he insists that we cannot pin down duration in a measurement. There is and can be no standard unit for duration. We do not measure duration when we measure time. Instead, we abstract from duration in order to construct a time series.

This point remains true even when we measure what psychologists and neuroscientists call "subjective duration," the length of time attached to a stimulus by a perceiver. This is clock time applied to perception, not duration in Bergson's sense of the term. Indeed, it would be wrong to think that Bergson's idea of duration can be assimilated to the idea of psychological time as opposed to physical time. Bergson is not saying that duration is a psychological phenomenon, whereas clock time is physically real. On the contrary, he asserts that time as defined in physics—clock time—cannot be detached from time as passage as given in the experience of duration. Clocks require clock readers, clock reading requires consciousness, and consciousness is inherently durational. These points will be important later in the book when we return to Bergson and his famous debate with Einstein about the nature of time.

Timekeeping

From sundials and water clocks to hour glasses and weight-driven mechanical clocks, timekeeping devices have a long and fascinating history, driven mostly by an unrelenting quest for better precision. Nevertheless, we should remember that the usefulness of any timekeeping device depends on our gathering information through our senses, usually by looking—the location of the shadow projected on a sundial, the position of the hands of a clock, the readings of the vibrational frequency of a quartz crystal—and thus relates directly to our lived experience. A timekeeping device translates the ineffable experience of passage into the language of numbers. In doing so, it appears to reify time, giving it a mathematical precision comparable to that of physical measurements, such as distance, weight, speed, or pressure. The more precise the clock, the more distant clock time seems to be from becoming, passage, and duration.

Today's timekeeping standard uses the frequency of specific electronic atomic transitions. Atoms have the advantage of bypassing irregularities in astronomical motions that require constant recalibration of clocks. The Swiss atomic clock FOCS-1, for example, relies on the frequency of electron jumps between energy levels in a cold cesium-133 atom, which is constant to very high accuracy so long as the clock remains in the same geographic location and at rest with respect to the ground. Under these conditions, the time interval of 1 second is currently defined as equal to 9,192,631,770 periods of orbital oscillations with this frequency. The FOCS-1 clock has the impressive uncertainty of 1 second in 30 million years.[9] The precision depends on the details of the experimental setup, consisting of a micro-wave cavity tuned to resonate with the electron jump frequency, like a parent pushing a child on a swing keeping the same period.

Still, no physical measurement can be absolutely precise. Every tool or device has an accuracy determined by its design. If a clock has a precision of a nanosecond (one-billionth of a second), it cannot be trusted to capture details of phenomena happening at scales of a picosecond (one-trillionth of a second). Thus, to every layer of reality based on measurements at a given scale, there is an underlying stratum that remains elusive. Even if, mathematically, we can divide time into smaller and smaller chunks, we could not expect to measure such shrinking time intervals ad infinitum. Infinitely divisible, physical time is a mathematical abstraction. The time line, a straight line covering the real numbers, is a useful tool for modeling time-varying phenomena with roots in the late Middle Ages (see chapter 2). It should not, however, be taken as representing the reality of time any more than a clock dial does.

Clocks don't reveal the true nature of time; they are tools invented to abstract certain aspects of the experiential flow of time and to measure them in a systematic way. Modern clocks are products of the scientific workshop, the collective effort of scientists and engineers to isolate aspects of experience and construct measurable invariants from them. But regardless of the precision of the clocks emerging from the workshop, our understanding of time remains rooted in duration, the irreducible experience of becoming.

Time in the Blind Spot

To consider the time of physics—what a clock measures—as the only real time is a clear example of the chain of thinking leading to the Blind Spot. First, we surreptitiously substitute mathematical time for lived time. Next, we commit the fallacy of misplaced concreteness by declaring that abstract, mathematical time is real time. Finally, we forget that the concrete being of passage as given in duration is the primal source and condition for the meaningfulness of the notion of time. This forgetting is the amnesia of experience.

The blind spot view of time leads to quandaries. Mathematically, as we consider ever shorter time intervals, what we call the experience of now-ness evaporates into no-duration. No-duration not only clashes with our immediate experience of time's passage, its ever-flowing nature, but also elevates mathematical singularities into puzzles and inconsistencies. How could something that lasts be constructed out of point-like moments, defined as instants with no duration? How seriously should we take scientific statements about the nature of physical reality that are based on a time interval that approaches zero? As we will see when we discuss cosmology (see chapter 5), these questions become acute when we try to make sense of the origin of the universe, when we try to answer, within the framework of physics and cosmology, Leibniz's famous question, "Why is there something rather than nothing?"

We need to distinguish the purpose of a specific notion of time from the impulse to attribute ontological primacy to it as a result of the amnesia of experience. To describe natural phenomena, the scientific narrative needs to abstract, to the largest extent possible, the passage of time from the human experience of duration. In science, the passage of time must be orderly and precise, the same for all observers with identical clocks, at least for those in the same reference frame (clock times differ for frames of reference that move relative to one another, according to the theory of relativity, but the theory also teaches us how to patch these differences). Physical time must have a universal standard, a demand that led to the "God-like" perspective of Newton's absolute time (see below). Nevertheless, the scientific need to use a mathematically accurate definition of time should not be confused with that definition's having any kind of ontological primacy. Insisting that it does was a major contributing factor to the Blind Spot of classical physics.

Human time includes lived time and abstracted mathematical time lines. The latter emerge from the former. We could not have built an abstract notion of physical time without first having the experience of time's passage. The mathematization of time through its representation on a continuous line composed of instants with no duration is a map, the passage of nature is the landscape, and our ineffable experience of time's flow—Bergson's duration—is the vehicle of our journey through the landscape. The map serves a clear purpose: the mathematical description of natural phenomena to the highest degree of precision possible. But you cannot be a mapmaker if you cannot see what you are mapping. The mapmaker should not forget what cannot be included in the map—the experience of walking the terrain, the biting chill of the mountaintop, the dappled light through the trees of the forest. Which details are important for which purpose? The map will get its users lost if the mapmaker does not understand its purpose.

Coarse-Graining

Before going further, we need to forestall an objection. Isn't our sense of passage in duration itself a cognitive construction arising from a measurable physical system, namely, the brain? Isn't duration the result of the brain's doing something like what physicists call "coarse-graining"—smoothing over fine, granular details by simplifying and integrating them? Isn't duration a result of smearing over granular clock time?

Consider a flowing river, an often-used analogy for the passage of time. From a macroscopic length scale, the river flows smoothly. Microscopically, however, a fluid disappears into a collection of molecules endowed with an orderly collective motion. That motion is certainly not smooth and flowing at short scales. The notion of a "fluid" is not applicable at the molecular scale. From a molecular perspective, to describe something as a fluid is to give a coarse-grained description. A granular system observed from a large enough distance appears fluid or continuous, like sand dunes seen from far away. Short-scale fluctuations are averaged away into a smooth continuum. This is one reason that the river of time is such a powerful image: it mirrors the coarse-graining of time that our perception performs, so that we experience what William James (following E. R. Clay) called "the specious present."[10] This is the "duration-block" that we perceive as "now" but contains within it a sense of the immediate past and an opening into the imminent future.

Recent research in cognitive neuroscience has reinforced the idea that the brain constructs the perception of the duration of the present moment by careful integration of asynchronous sensory and motor activities. Since sensory stimuli, such as sound and light, travel at different speeds, as do exteroceptive and interoceptive signals along nerve pathways, every event that we perceive as happening "now" reflects a cognitive performance by the brain (though there is little consensus about exactly how this happens).[11] As James already knew, using clock time to measure the specious present gives highly variable results, depending on a host of factors, such as emotion, mood, motivation, arousal, feeling states of the body, ongoing endogenous neural rhythms, and which sensory system is being investigated.[12] In any case, the variable specious present reflects the brain's coarse-graining of spatially and temporally distributed neuronal activities at multiple scales.

Fascinating as this research is, it is important to realize that it proceeds by applying clock time to the outwardly available neurophysiological and behavioral accompaniments of inwardly felt duration. The research investigates the "neural correlates" of the conscious experience of duration. Although it provides valuable information about how the brain parses time, it does not explain duration in the sense of showing how brain activity—described strictly from the outside in third-person terms—suffices for the subjective experience of duration. To think that neuroscience does give such an explanation—that we can understand duration entirely in terms of the brain's coarse-graining activity—implies that neuroscience has solved the problem of consciousness, which is manifestly not the case (see chapter 8).

What is really going on is that scientists, relying on their own first-person experiences of duration, apply clock time to observable biological and behavioral processes, and then infer back to aspects of duration that they can extract and stabilize as objects of thought and attention, and describe in intersubjectively agreed-on ways. They never get outside duration and explain it in terms of something else. Indeed, thinking you can get outside duration and explain it as resulting from the coarse-graining of minuscule, granular units of clock time is incoherent because clock time has no meaning without duration.

As Whitehead clearly saw, there can be no explanation of passage, because every explanation, particularly in terms of clock time, presupposes the passage of nature given in duration: "Nature is a process. As in the case of everything directly exhibited in sense-awareness, there can be no explanation of this characteristic of nature. All that can be done is to use language which may speculatively demonstrate it [i.e., point to it so we see it], and also to express the relation of this factor in nature to other factors."[13] In other words, passage as given in duration is primitive (basic and underivable). We can abstract from passage with amazing technical precision, but we can never reconstitute it out of those abstractions or replace it with them. We will see this point illustrated repeatedly in the rest of this chapter (and in chapter 5).

Newton's Absolute Time

We now turn to the mathematization of time in physics and how it is responsible for some of the paradoxes we face concerning the nature of time. In the previous chapter, we traced the ascending spirals of abstraction that led from the Mertonians through Newton and on to Hamilton in the development of mathematical laws of motion. Here we emphasize the essential role that time played in those developments. Extracting the proper abstractions to create laws of motion meant creating a proper, or at least useful, abstraction for time itself.

Recall that if we assume that mathematics is the language of nature, as Galileo once remarked, we must first universalize time in order to chart change in the state of motion of an object or objects (motion being the embodiment of change in natural systems). This means that time must pass everywhere in the cosmos, even if we can't sense this directly. This apparently obvious statement is a very recent invention. For Aristotle, and natural philosophers for over two thousand years after him, the heavens were essentially timeless. The cosmos was uncreated and celestial objects circled about the Earth carried by crystal spheres. There was motion in the skies, since planets were seen to move with respect to the constellations, but it was motion with no change. The heavenly bodies were not in time because there was no "before" and "after" for their motion, and their total motion could not be measured by time, even though they stood in certain temporal relations, such as simultaneity, with sublunary things (they circle the Earth while things are happening on it).[14] Aristotle and his followers split the cosmos into two realms, celestial and terrestrial, each obeying a different set of rules. In particular, change was allowed only in the terrestrial, or sublunary, realm. This binary split served the purposes of the Church quite well: change and decay were relegated to Earth and life (the realm of becoming) and changeless bliss to the heavens (the realm of being). The first cracks in the Aristotelian worldview were exposed in the sixteenth century thanks to Tycho Brahe and his sightings of the 1572 supernova and the Great Comet of 1577, both of which he determined were located beyond the sphere of the moon. From these observations it became clear that change, and time, also flowed in the heavens. The skies had a history.

We have seen that during the fourteenth century, scholars at Oxford University's Merton College implemented the idea of time as a measure of change to study motion. After slow beginnings, their groundbreaking mathematical representation of time would finally take hold during the early seventeenth century. Two of Kepler's three laws of planetary motion refer explicitly to time. His second law stated that the line connecting the sun to an orbiting planet sweeps out equal areas in equal times. His third law equated the square of a planet's orbital period to the cube of its mean distance from the sun. In Italy at about the same time, Galileo obtained the law of pendular motion relating the period of oscillation of a pendulum to the square root of its length. Galileo also recognized that free-falling objects cover a distance proportional to the square of the time, obtaining the law of free fall. In this way, through the work of Kepler and Galileo, time came to play a central role in the earliest mathematical laws describing terrestrial and celestial natural phenomena.[15] About half a century later, Newton would unify Galileo's law of free fall and Kepler's laws of planetary motion as different applications of the same principle, the law of universal gravity. Time, Newton realized, was the thread weaving terrestrial and celestial physics into a single tapestry.[16] This profound innovation allowed for the development of a scientific narrative of natural phenomena that embraced the whole universe, rendering it all accessible to the human mind.

Newton's unification of terrestrial and celestial physics called for the existence of one time, running its course across the vastness of space. The equation that tells how the apple falls to the ground or how the moon orbits the Earth uses the same time variable t. This led Newton to propose a definition of *absolute time*, the cornerstone of his formulation of the laws of motion, a compromise between our lived experience of time as a continuous flow from the past to the future and a rigid mathematical framework:

> Absolute, true, and mathematical time, of itself, and from its own nature, flows equably without relation to anything external, and by another name is called duration: relative, apparent, and common time, is some sensible and external (whether accurate or unequable) measure of duration by the means of motion, which is commonly used instead of true time; such as an hour, a day, a month, a year.[17]

Following Galileo's lead, Newton realized that an observer in inertial motion—motion at a constant speed and on a straight line—could not differentiate this motion from being at rest. We know this from experience. Imagine traveling in a car at 60 miles per hour on a straight line with your eyes closed and ears plugged. Unless you open your eyes and look outside,

you can't tell whether you are moving or at rest. Motion is always in relation to something else. Thus, the study of motion wasn't merely concerned
with changes in position from point A to point B. After all, we can't really
do that at constant speeds. (To achieve a constant speed, you need either to
accelerate from rest or from another speed.) Indeed, such inertial motions
are really an abstraction rather than a concrete reality. (They are one of
the unreal idealizations discussed in chapter 1.) The secret, Newton understood, lies with the agent that changes an object's state of motion, that is,
that imparts the object with an acceleration. As we discussed in the previous chapter, this agent is a force, capable of changing an object's momentum, $F = d\,(mv)/dt$.

Newton's second law tells us something remarkable about time in classical physics. Since velocity is the change in the position of a body over a
time interval ($v = dx/dt$), it follows that a change in the state of motion—the
acceleration caused by a force—is a change in time of the velocity (keeping the mass constant for simplicity): $a = dv/dt = d(dx/dt)/dt = d^2x/dt^2$. The
acceleration is the change of a change (of position), and thus second order
or quadratic in time. This is why we see velocity in units of, say, km/h and
acceleration in units of km/h^2. (More commonly, time appears in units of
seconds, not hours.)

What's remarkable here is that since time appears squared in the equation, if we reverse the direction of time flow, that is, if we go from present
to past (just change the sign in the time variable to $-t$ in the equation),
the equation remains the same. (Recall that the square of á negative number is positive: $(-1)^2 = -1 \times -1 = 1$.) In other words, Newtonian mechanics
is time reversible. It doesn't distinguish time running forward from time
running backward. This remains true even for the more sophisticated post-
Newtonian formulations of classical physics based on the action principle
discussed in the previous chapter.

Picture a movie of a ball moving about in space from left to right. Of
course, this is a highly idealized situation, and it is important to keep this in
mind. Still, playing the movie backward, that is, reversing the direction of
time, you'd see the ball retracing its motion from right to left. Time-reversal
invariance simply means that if you didn't know the original direction of
the motion, you couldn't tell which way was forward or backward in time.

In other words, although Newton's absolute time flows, it could equally
be flowing to the future or to the past. It is therefore not like a river at

all. Rivers always flow downward due to gravity. In the highly idealized physical situations where there is no loss of energy due to inelastic collisions (friction), the abstracted time that is central to the abstracted laws of physics can move forward or backward. In this sense, the time of physics is utterly unlike the time of our experience.

Initial Conditions and Representations: How Physics Tells Stories

How could the equations that describe motion not distinguish between past and future? It's an idea that's not only puzzling but perhaps even disappointing. If true, how could these exalted equations of physics be of any use to describe a reality where the flow of time, what is past and what is future, is obvious to our experience?

Here, we must pause briefly to discuss what these equations represent and how they are deployed. Equations in physics model idealized situations. They don't describe things as they are in themselves—an impossibility anyway, as we can't know things as they are in themselves—but rather phenomena mapped through our observations and measurements, simplified to ignore effects that are believed to be insignificant. In other words, equations are abstract constructions, translations from the world of perception to the world of reason: they originate in experience but exist in thought.

"We must observe the immediate occasion, and *use reason* to elicit a general description of its nature," wrote Whitehead (his emphasis).[18] Equations are attempts to encapsulate the characteristics of a physical system with the goal of describing its observed behavior, usually its evolution in time. This encapsulation identifies the structural invariants of observable behavior, which serve as the basis for having any natural science at all. Thus, a successful physical model is one with equations that, once solved, describe the observations of the system at different times. For example, we can use Newton's law of universal gravity to predict when and where the next total solar eclipse will occur with remarkable accuracy. Note that for the particular case of planetary dynamics, it doesn't really matter whether the celestial objects are moving clockwise or counterclockwise. Planetary systems are essentially friction free, making them ideal subjects to model with Newtonian mechanics or other action-based formulations of classical dynamics. This modeling works, however, only if we ignore these systems' turbulent early formation history as well as the effects of tidal friction.[19] Friction, also known as "dissipation," breaks the symmetry between time flowing toward the past or toward the future. Prediction and postdiction are possible and indistinguishable only in the rare classes of natural phenomena efficiently mapped by friction-free models. This is one of the reasons that the descriptions of celestial orbits were among the first applications of mathematics to natural phenomena, shaping the evolution of physics for about two centuries. Still, historical contingency should not be a criterion for universal validity. Science is often shaped by the kinds of questions it can answer, not by the kinds of questions it should.

To solve an equation representing a physical system, we need to specify both the position and velocity of the objects comprising the system at the beginning, at $t = 0$. These are called the initial conditions. The same is true for Lagrange's and Hamilton's action formulation of classical mechanics, where the time interval is bracketed between the initial and final time when motion takes place. Every equation tells a story, and the story must have a beginning. Here, the "beginning" is simply the instant in time denoting when the model is supposed to start tracking the motions under study or when the experiment began. It is an abstract point in the abstract time line.

The "objects," often idealized as point particles of mass m, could be planets orbiting the sun, idealized electrons flowing along a wire, the molecules of a gas confined to a room, a ball rolling down an inclined plane, the bob of a pendulum—whatever system the model is attempting to describe. By idealizing the objects as structureless particles, the modeler assumes that their shapes or sizes are irrelevant to the dynamics.[20] For example, when modeling the orbit of Jupiter around the sun with Newtonian mechanics, we don't need to know the details of Jupiter's gaseous composition or the nuclear reactions that fuel the sun. All we need to know are the values of their masses and their mutual distance. Physics is the art of efficient modeling. As long as the approximations hold, the model is reliably predictive. That Newton could describe so much in his time—the precession of the equinoxes, the tides, the slightly oblate shape of the Earth, the orbits of planets, moons, and comets—is a testament to the spectacular success of his approach to motion. Inevitably, that success meant that the Enlightenment's rationalization of nature cemented the blind spot view of time as an essential aspect of physical reality.

The worldview of classical physics is an idealization of reality that relies on a mathematical, abstract notion of time that ticks away with absolute precision. Causality is assumed implicitly: time is ordered. Equations track effects that are due to specific causes. To solve the equations modeling how a particular system evolves in time, one must feed in the initial conditions. Where are the particles at the beginning of the motion? How fast and in what directions are they moving? To tell the story of motion, the modeler needs to know the numbers denoting the initial positions and velocities. This is crucial.

But how are these numbers known? They are known because they are measured. And who measures these numbers? The observer does, either directly or through an apparatus. An apparatus, that essential product of the scientific workshop, is understood as a measuring device that provides sensorial information to the observer: clicks, numbers on dials and screens, blips, graphs, and visual patterns. It is clear that experience lies at the heart of every story we tell about motion. Since we humans are the ones telling this story and taking the measurements, we can know the numbers representing initial conditions only to a certain precision.

When we say that the ball is held at rest 1 meter above the ground at $t = 0$, to what precision do we know this fact? To how many decimal places do we know the ball's initial height H? $H = 1.00001$ meter? $H = 1.001$ meter? $H = 0.999999$ meter? Does high precision matter? Not always, and this is why this approach is so successful in many applications. In particular, high accuracy in initial conditions doesn't matter much for linear systems, defined as systems that respond to an impulse in direct proportion to the strength of that impulse. In the absence of friction, linear systems are the ones that are time reversible. They are systems where time can flow in either direction and the behavior would look the same. Unfortunately, such systems are few and far between.

Most physical systems in nature are nonlinear. That means they can exhibit chaotic behavior, with outcomes that are extremely sensitive to initial conditions. Make little changes in the precision of these numbers and the equations will tell a very different story. For chaotic nonlinear systems, the future becomes an unfathomable vastness of possibilities, springing from minute—and often unknowable—differences in the numbers that describe the initial conditions. In such cases, one can't tell the story backward in time, a point stressed by physicist George Ellis.[21] Complexity leads to a loss of time-reversal invariance, a point we will soon explore in more detail. More generally, and this includes both linear and nonlinear systems, there is an unquestioned assumption that every physical quantity, including the numbers denoting initial conditions, has a "real" numerical value of infinite precision, thus falsely attributing physical reality to the numbers that denote such quantities.[22] This assumption is a fantasy: nothing that humans can know can be known to infinite precision. In every practical application, numbers denoting physical quantities are always truncated to a certain number of decimal places. It follows that given the irreducible centrality of experience in the scientific narrative of the world, epistemology frames ontology: what we can know of the world determines what we can say about its physical nature. This is what is lost in the Blind Spot of classical physics.

Infinity is another case in point. The infinitely large or infinitely small is an idea, not a number. It is extremely useful mathematically and gets invoked as a tool in many key proofs. Most famously, infinitely small increments are at the heart of calculus, and when the increments are of time, at the heart of all the equations of motion in physics. But we could never be certain that anything in physical reality is infinite in any way, since "infinitely small" and "infinitely large" lie forever beyond what we can measure. As the mathematician David Hilbert once stated, "The infinite is nowhere to be found in reality, no matter what experiences, observations, and knowledge are appealed to."[23]

Experience is at the very heart of every physical model and every statement we make about perceived reality, placing a wedge between what we can know of the world and the false hope that science can provide a God's-eye objective view of physical reality. Experienced time is modeled as mathematical time in the equations of physics. But time is not made of numbers, and certainly not of numbers of infinite precision. Time is constructed out of passage, and passage is known only through experience, the experience of duration. Experienced time is the only time we know. The reification of mathematical time is an unjustified extrapolation from model to reality, a dangerous Blind Spot confusion between map and territory.

Thermodynamics and the Arrow of Time

The story our experience of time's passage tells is very different from the time reversibility of friction-free, idealized Newtonian and post-Newtonian mechanics. Things fall and break down, sugar dissolves in coffee, waves crash at the shoreline, rain comes down from clouds, caterpillars turn into butterflies, we age, we die. None of these events happens backward. Our experience of reality has a clear direction in time, from past to future. This is how our experience of being-in-the-world manifests itself. Even when we attempt to shut off any connection with the outer world, be it in deep meditation or in a sensory deprivation tank, our hearts are still beating and our blood is still pulsing through our veins. Our mind still senses an ongoing torrent of thoughts, flowing from some unknown source with a consistent direction. This source is part of the wellspring of being alive and existing in time: memories of the past accumulate as we experience the present. We don't remember the future. It's clear that our experience of time is at odds with the mathematical time of the scientific narrative.

Physicists in the late nineteenth century knew they needed an account of how time appears to move forward. They needed to show how a direction for time could emerge from the directionless variable t in their dynamical equations. As we will now explore, the path they took trying to find this reconciliation began not in some lofty theoretical framework or metaphysical insight, but at least initially, in the practical engineering concerns of improving the fuel efficiency of machinery that uses heat to generate motion. Once again, the spiral of abstraction and explanation began in the workshop.

During much of the nineteenth century, the Industrial Revolution triggered interest in the workings of steam engines, used in transportation, mining, and farming. The more efficient the engine, the more useful work it could perform with less fuel consumption: efficient engines increased the accumulation of capital, providing one form of the critical, entangling links between science and industrial economies that we explore in the final chapter. This financial motivation opened a new playing field for physics: that of the study of heat and heat transfer. During the transition from the eighteenth to the nineteenth century, the nature of heat changed from being a mysterious substance that could flow from hot to cold objects (first "phlogiston," then "caloric") to the understanding that heat is not a substance at all but motion. It is the thermal agitation of small particulates of matter, often invisible to the naked eye.[24]

With the study of heat, the atomistic nature of matter reentered physics after a long hiatus. Daniel Bernoulli, John James Waterson, and James Clerk Maxwell all showed how gas molecules could be modeled as elastic spheres (no loss of energy in collisions). In this way, they obtained mathematical expressions relating macroscopic properties in a gas, like pressure and temperature, to the average molecular velocity and to its square, respectively.

We know from experience that if we pour a bucket of hot water into a tub filled with cold water, the water in the tub will reach some intermediate warm temperature. The higher the volume ratio of hot to cold water, the higher the final equilibrium temperature of the tub water. Microscopically, this temperature increase results from the energy transferred from hot (fast) water molecules colliding with cold (slow) water molecules, leading to a higher average molecular velocity in the tub, and thus to a higher overall water temperature. We also know, from experience, that cold water doesn't turn hot again unless heated by a source. Microscopically, this means that on average, molecules cannot start moving faster on their own. Thus, the direction of time is hidden in a phenomenon as simple as the cooling of water.

Note how we used the word *average* when characterizing the velocity of molecules above. The emergence of atomistic thinking in the description of macroscopic properties of gases (including steam, of course) imported a new mathematical element into physical models: statistical analysis. Although the microscopic mechanics of particles colliding with one another is still deterministic, it became impractical or even impossible in principle to describe the behavior of a gas made of many particles by jointly solving the equations of motion for each one of them.

Recall that to solve each equation, we would need to know the initial position and momentum of each of the molecules with a certain precision. As an illustration, 1 mole of a substance contains about 6×10^{23} particles, which can be atoms, molecules, ions, or electrons. That's a 6 followed by 23 zeros. (As an illustration, the average mass of 1 mole of water is 18.02 grams.) To solve the equation for each water molecule, one would need to specify three numbers for the initial position and three for the initial momenta, the six "degrees of freedom" of a particle moving in phase space.[25] So, for a number N of particles, we would need to solve $6N$ equations. The largest numerical simulations that can be solved on a supercomputer at present can include about 10 billion particles, or $N = 10^{10}$. Although impressive, this number is thirteen orders of magnitude smaller than the real situation. Clearly, statistical and approximation methods are required.

Recall the image of a bucket of hot water thrown into a bathtub filled with cold water. Macroscopically, we can measure the change in the water temperature with a thermometer. The larger the amount of hot water poured in, the higher the average temperature becomes. This we know intuitively, and thermometer readings will confirm that intuition, since, as we have seen, temperature itself was built from these kinds of experiences. We also know that if we mix the water, eventually the whole bathtub will reach a higher temperature than its initial value and then remain steady (ignoring cooling with air). This approximate stasis, where the water temperature remains constant, is a state of thermodynamic equilibrium. If the bathtub were a perfect closed system, unable to exchange energy with the outside, the water temperature would remain constant forever. Viewed macroscopically, thermodynamic equilibrium is timeless.

Microscopically, the situation is very different. Water molecules are whizzing by, colliding with one another at varying speeds. Locally, there is a lot of activity. The beauty of statistical averaging is that it validates the mathematical representation of our $N = 10^{23}$ particles in their $6N$-dimensional phase space. The averaging allows the abstract map of phase space to make contact with the terrain of our actual experience of ice melting and bathtubs filling with hot water. It lets us focus on what we believe matters in our experience (its structural invariants) in making an accurate map. Once we sum over all the molecules and average over the possible variations in speeds, the individual details of this very complicated, local, invisible, and abstracted microscopic dynamics are ironed out. For individual molecules in a gas in thermodynamic equilibrium, time is very much present. But for outside macroscopic observers, it is not.

From this example, we also see how macroscopic properties such as pressure and temperature are emergent behaviors: they make sense only when we look at a system—or a sufficiently large portion of it—holistically. We can't speak of temperature for one or a few molecules.

This emergent approach is essential for the thermodynamic interpretation of time's arrow. In 1872, Ludwig Boltzmann formulated his famous "H-theorem," demonstrating that an efficient way to characterize the behavior of a thermodynamic system is through the identification of how its specific states evolve in time. In this view, the state of the system is a sort of histogram portrait of the positions and momenta of its individual constituents. Molecules (or atoms, ions, or whatever the system's constituents are) are binned according to where they are in space and the values of their momenta, giving us the six degrees of freedom we mentioned earlier.

As the system evolves in time, molecules will collide and change position and momentum, and the binning will also change. Some fast-moving molecules may become slow moving and vice versa. Thus, we can think of the evolution of the system as a series of portraits of the relative binning of positions and momenta in time, the phase space we described in the previous chapter.

In our example of the bucket of hot water poured into a bathtub of cold water, the initial state would have two bins dominating the histogram. One bin holds the hot molecules of the bucket (high temperature meaning high average velocity). The other bin holds the lower velocity (lower temperature) molecules in the colder bathtub water. This second bin has far more molecules than the one representing the bucket because it represents more water. As time passes, collisions between hot and cold molecules would reduce the relative velocities until eventually, most molecules would have approximately the same average warm velocity with only small fluctuations about the mean value.[26] This is the state of thermal equilibrium. In phase space, a single bin would dominate the distribution, characterized by an "equipartition of energy," meaning a state with the probability of finding a molecule with (or very near) the average velocity being the highest.

Equipartition of energy, the ultimate molecular democracy, is thus equivalent to thermodynamic equilibrium: the water in the bathtub with a warm final temperature. From then on, at least for a closed system, no significant macroscopic change would be detected. In particular, we wouldn't see the system spontaneously reverting to its initial state, with hot and cold water separated in space.

Boltzmann's celebrated H-theorem attempted to provide a quantitative representation of how reversible microscopic mechanics (the molecules individually colliding with one another) can explain macroscopic behavior (the water eventually reaching a warm equilibrium temperature). As the system evolves in time, Boltzmann showed that a quantity H (the inverse of what we call the entropy of the system) moves toward a minimum value that characterizes thermal equilibrium. To revert a system that has reached equilibrium to its initial nonequilibrium state would necessitate an absurdly large number of collisions with a very specific mandate, amounting to a phase space evolution with a negligibly low probability of occurrence. According to Boltzmann's H-theorem, for systems with many constituents, and thus a huge number of possible states, going back in time is not impossible but highly improbable. With the help of probability, Boltzmann concluded that the arrow of time points in the direction of equipartition of energy, the state with maximum entropy.

At the time, Boltzmann believed he solved the problem of the arrow of time. He thought he had translated the physics of microstates into the time of experience with its movement from past to future. But other physicists soon raised important questions about Boltzmann's approach. The conflict between the two points of view reveals another expression of the Blind Spot in the history of physics.

The Reversibility Paradox

Just a few years after Boltzmann proposed his theorem, one of his colleagues from Vienna, Josef Loschmidt, argued that one cannot ever prove conclusively time irreversibility starting from time-reversible equations. This "reversibility objection" became known as Loschmidt's paradox (also called the reversibility paradox). The key idea is that it is possible, in principle, for a rare state to occur where the velocities of the molecules at some time t along their evolution are reversed in such a way as to force the system to backtrack to its initial, more ordered state. This would mean a decrease in entropy toward the future. For long enough times, such states do have a finite probability of appearing and will occur even if they are highly unlikely to show up over short times.

The roots of Loschmidt's paradox are hidden in the Blind Spot, in the insistence that we have, at least in theory, unlimited access to information and detail about measurements of time and other physical variables. In his classic book *The Direction of Time*, the physicist and philosopher Hans Reichenbach argued that even if we considered a system with indeterministic equations or one that was not strictly closed, Loschmidt's reversibility objection still holds: "All that can be proved in this way, however, is that the reverse process is made even more improbable, not that it is excluded."[27] In other words, we can always observe the system for arbitrarily long timescales, and at those long times, an occurrence of a state that effectively decreases the entropy, or even reverts the system to its initial, low-entropy state, will occur. This perspective on Loschmidt's paradox is often called Zermelo's recurrence objection, after Ernst Zermelo applied Poincaré's work on the recurrence of phase space trajectories to the H-theorem in 1896.

There is a problem, however, with both Loschmidt's paradox and Zermelo's critique, which hovers not only over their own work, but also over Boltzmann's entire project of rescuing the arrow of time from time-reversible microscopic motion. The problem lies in the definition of the *state of the system*. As we discussed above, when describing the behavior of systems with many particles, one must use the phase space description, whereby one groups together particles with close enough position **x** and velocity **v** (or, equivalently, momentum **p**). (Bold characters indicate that these quantities have three entries each [they are vectors], since particles are moving in three dimensions.) For N particles, a small region (or "cell") of phase space would have volume $(\Delta\mathbf{x})^N(\Delta\mathbf{p})^N$. What size do these cells have? That depends on the resolution of the measuring devices, which will *always be limited*. This practical precision limit clashes directly with the assumption of infinite precision measurements. There is, therefore, a tension between the way we interact with natural systems, which is always coarse-grained, and the assumed infinite precision of some of our models.

To understand why this is an issue, picture two particles colliding, like billiard balls. We assume that before colliding, they travel independently from different locations, representing what we call an uncorrelated initial state. But once they hit each other, the details of the collision are imparted in their subsequent behavior. If it was a grazing collision, the two particles would behave quite differently than if they collided head-on. This means that after the collision, the two particles are *related* or, in physics terminology, they are correlated. A purely mechanistic reconstruction of the motions needed to revert the state of the system (and hence decrease the entropy) would require arbitrarily high resolution in phase space, that is, arbitrarily small cells in position and momenta. But there is always a precision limit to what we can measure, and this precision sets the size for the smaller cell. As a consequence, phase space is *necessarily* coarse-grained. (And we are not even talking about quantum indeterminacy, which sets an ultimate coarse-graining limit, as we discuss in the next chapter.[28])

It's here that the critical issue emerges. Due to the finiteness of the phase space cells, some information about the collision is always lost. It becomes impossible to reconstruct the correlations in all detail. In other words, every phase space description entails memory loss. Coarse-graining of phase space guarantees that the system exhibits what is called *molecular chaos*, which means that initial and final states remain uncorrelated. Only in this way can closed thermodynamic systems satisfy Boltzmann's H-theorem and thus distinguish between past and future.

Thus, it is the time asymmetry implicit in a coarse-grained description of phase space that would rescue Boltzmann's argument for the arrow of time from Loschmidt and Zermelo. The idea would be that deep in the microworld of a small number of particles, time has no definite direction. Time's arrow emerges as the number of objects grows, collisions increase in complexity, cells in phase space coarsen, and entropy increases. Time's direction would now appear as a consequence of our blurry perception of reality, a point also made by Carlo Rovelli.[29]

This rescue of the arrow of time, however, comes at a profound cost to the Blind Spot view of the universe. If our physical theories, models, and interactions with nature are, *and must always be*, coarse-grained and time's arrow emerges from coarse graining, then time's direction now reappears in physics as a consequence of the essential dependence of physics on our durational perception and narrative of reality. We are back to the impossibility of pulling apart physical time and lived time. We are back to human time.

The problem of time's arrow, viewed from a microphysical perspective, reveals that the perfectly omniscient "Gods-eye view" of classical physics was never more than a useful abstraction. It was a map that could not, even in principle, recover the terrain of lived experience.

But the microworld is not the only stage on which the problem of time's Blind Spot would appear in physics. The setup of closed thermodynamic systems modeled with colliding elastic hard spheres is too restrictive to encompass the huge variety of ways in which time flows forward independent of our models and observations. What of stars born in gas nebulae billions of years ago, of life unfolding on primal Earth, of waves crashing on lonely beaches, of the beating hearts of hummingbirds, of bacteria moving toward food? Nature knows nothing of phase spaces or coarse graining. It knows nothing of our idealized abstractions. Yet time has been advancing since the Big Bang. How come?

To answer, we must explore how Einstein's theory of relativity localized the flow of time, taking away its Newtonian universality. As we will see, every small region of space has its own time, ticking away according to the gravitational field in its surroundings and to the state of motion of the observer. This discussion will also bring us back to Bergson and his debate with Einstein about the nature of time, which took place in Paris in 1922.

But before we explore the early history of the universe, where the physics of space and time collides with the physics of the very small, we must first explore our current description of matter at the subatomic level. There we find discoveries that made the limiting views of the Blind Spot painfully apparent and forced a steep price for the maintenance of its prerogatives, a price many are unwilling to pay.

4   Matter

Philosophical Prologue

In light of our journey from the Greeks to the peak achievements of classical physics, and our consideration of the role of time in physics, it is worth pausing to consider again the metaphysical perspective that came to be attached to science and that we have identified as the Blind Spot. To do so we will draw from Whitehead's classic critique of the philosophical worldview that emerged from the early centuries of modern scientific progress.[1]

Central to that scientific worldview was the distinction, drawn by Galileo, Descartes, Boyle, and Locke, between primary and secondary qualities. Primary qualities are the fundamental qualities belonging to the minute particles of matter (corpuscles) whose relations in space and time constitute the order of nature. For Locke, these qualities were size, shape, motion, number, and impenetrability.[2] Secondary qualities (color, taste, sound, smell, and hot and cold) existed only as effects in the mind caused by matter's primary qualities. (Strictly speaking, secondary qualities are the powers of material objects, based entirely on their primary qualities, to cause mental sensations that do not "resemble" the primary qualities.) The primary/secondary quality distinction inaugurated what Whitehead called the bifurcation of nature. Here is his description of the upshot of that split:

> Thus the bodies are perceived as with qualities which in reality do not belong to them, qualities which are in fact purely the offspring of the mind. Thus nature gets credit which should in truth be reserved for ourselves: the rose for its scent: the nightingale for his song: and the sun for his radiance. The poets are entirely mistaken. They should address their lyrics to themselves, and should turn them into odes of self-congratulation on the excellency of the human mind. Nature is a dull affair, soundless, scentless, colourless; merely the hurrying of material, endlessly, meaninglessly.[3]

Although the conception of nature's primary qualities had evolved by the end of the nineteenth century, the triumph of classical physics had instilled the bifurcation of nature in the collective scientific culture. It was accepted that the world was made only of matter in motion and fields (such as electricity and magnetism). All macroscopic arrangements of matter could be reduced to the arrangements of tiny atoms that were both the source of gravitational and electromagnetic fields and responded to their presence. In many cases, the behavior of those atoms (their motions and collective configurations) could be described with highly abstract mathematical structures and the techniques developed for their manipulation, including approximations when needed. In their most powerful and predictive forms, these mathematical structures cast the behavior of matter and associated fields into hyperdimensional "spaces" (that is, phase space) where motion and configuration manifested as the geometry of objects within those spaces.

We human beings and our experiences were on the other side of the bifurcation. Under the metaphysics attached to classical mechanics, experience was demoted to "nothing but" a world of appearances. The sun as "brilliant," the rose as "sweetly fragrant," the songbird as "mellifluous" were nothing but epiphenomena. They could be explained, in principle if not in practice, through their ultimate reduction to the behavior of atoms in brains (though how to bridge from the brain to mind remained a mystery). That behavior fell entirely under the jurisdiction of classical physics. Most important, for the aesthetic appeal of this metaphysics, all atomic behavior could be expressed elegantly and completely in the Platonic, gossamer mathematical constructions of a Lagrangian or the properties of a many-dimensional torus in Hamiltonian phase space.

This reduction or subtraction of experience into mathematical abstractions is exactly what Whitehead argued was a profound mistake and we argue is the hallmark of the Blind Spot. It's not that Whitehead opposed abstractions. He was a mathematician and logician, and he knew full well the glory and power of Lagrange's and Hamilton's achievements. His philosophical work, however, was an attempt to find the right place for such potent abstractions within a full account of nature that includes our experience as an integral part of it. Particularly important to us, in accounting for the limitations of the blind spot worldview, is Whitehead's critique of the use of abstractions. As he writes at the outset of *Science and the Modern*

World, "Philosophy, in one of its functions, is the critic of cosmologies."[4] By "cosmologies," Whitehead is referring to metaphysical views about reality and the kinds of things that make up nature, particularly those views attached to modern science. Philosophy can and should critique the often-unspoken philosophical background of science. Thus, when Whitehead calls attention to the "fallacy of misplaced concreteness," the "error of mistaking the abstract for the concrete," he means to highlight how mathematical abstractions have come to be reified and mistakenly seen as both replacements for and explanations of what we experience.

By the beginning of the twentieth century, the enormous success of classical physics appeared to justify giving its abstractions pride of place in an account of the world. In that world, lived experience was only a "just"—just sensations, just psychological states, just epiphenomena. Classical physics provided the cosmology, the worldview, in which the basic ontology of nature mattered most. Classical physics gave an account of the furniture of the universe. Its job was to tell us in exacting detail what really existed out there in the world, independent of us humans and our messy biologically based and psychologically mediated experience. Physics would unveil the fundamental structures and their relations. In doing so, it would account for all that exists now and forever. Its domain was reality seen from an external, God's-eye point of view.

The success of classical physics led to the belief that those basal structures and their relations, now represented by atoms, fields, and their mathematical laws, had been fully enumerated and articulated. All the rest, including experience, could be recovered through the proper procedure of reducing the world's apparent complexity to those basal forms. Physics gave us the ultimate ontology of nature.

As Whitehead and Husserl saw, however, this way of understanding what physics had achieved was misguided. It confused abstract idealizations with concrete being and its touchstone in bodily experience. As Whitehead wrote, "Thought is abstract; and the intolerant use of abstractions is the major vice of the intellect."[5] For Husserl, this intolerant use is the surreptitious substitution, the substitution of the abstract for the concrete.

Our tour of classical mechanics and analysis of the problem of time illustrates how the construction of classical physics, like the invention of temperature, represented an upward spiral of abstractions that nonetheless always relied on embodied experience for its source and support. While this upward spiral was grounded on the importance given to mathematics in the Pythagorean and Platonic tradition, it took its first true steps when the Mertonians found a way to extract time and motion from experience and reformulate them in terms of the abstract concepts of velocity and acceleration. Later, after Galileo, the scientific workshop and its communal activity became the essential new social arrangement allowing progress in physics to accumulate. Through the workshop, a collective effort was undertaken to isolate exactly those aspects of experience amenable to serving as structural invariants for abstraction and idealization. This is how the upward spiral of abstraction accelerated, taking us from Newton's laws to the far more generalized coordinates of Lagrangian and Hamiltonian mechanics.

The role of embodied experience, however, was never lost. Consider the principle of least action (or stationary-action principle): when a particle of fixed energy travels from one point to another, its trajectory is such that the corresponding action has the minimum possible value. Although the mathematical derivation and statement of the principle may seem the pinnacle of abstraction, anyone who has hiked up a steep, obstacle-filled mountain trail quickly becomes intimate with the idea that nature settles on some form of minimization in its expression. There is a rhythm the body falls into after an hour or two of effort. Like a dance, each step is chosen to maintain the motion upward with an economy of effort. The body, in other words, knows. Just as the Mertonians had to transform embodied experiences of "quickness," "rapidity," and "fastness" into a usable mathematical form (velocity and acceleration), so the principle of least action did not spring forth fully formed like Athena from Zeus's head. Discovering the principle would not have been possible without the concrete antecedents of experience in the life-world. We must live the world before we conceptualize it.

Thus, we come to Whitehead's insight concerning the relationship between experience (which he refers to as the concrete) and the abstractions of science, as seen from the perspective of philosophy as the critic of abstractions:

> The explanatory purpose of philosophy is often misunderstood. Its business is to explain the emergence of the more abstract things from the more concrete things. It is a complete mistake to ask how [the] concrete particular fact can be built out of universals. The answer is, "In no way." The true philosophic question is, How can concrete fact exhibit entities abstract from itself and yet participated in by its own nature?
>
> In other words, philosophy is explanatory of abstraction not of concreteness.[6]

Thus, the blind spot metaphysics misunderstood its own purpose. It reduced concrete experience to scientific abstractions, or replaced one with the other. We have argued that this is a grave mistake. It has taken science into erroneous problems and taken the culture on which science rests into existentially dangerous situations. We discussed the problems concerning time in the previous chapter, and we return to them in the next chapter on cosmology. The dangerous situations will loom large when we discuss AI (chapter 7) and the climate crisis (chapter 9). In this chapter, we stay within the framework of physics. Nothing illustrates the Blind Spot as dramatically as the emergence of quantum physics during the first three decades of the twentieth century. It was then that physicists faced, often with desperation, the unavoidable crumbling of the mighty classical physics worldview.

The Workshop Triumphant: The Birth of Quantum Physics

Nobody asked for quantum mechanics, and in a sense, nobody really wanted it when it arrived. The domain of physics that resides behind the word *quantum* was forced on physicists as a direct consequence of the workshop: scientific technology made us do it.

In the late nineteenth and early twentieth centuries, technical advances in the form of new machines—new devices—allowed physicists to begin probing length scales and timescales far smaller than had been possible before. These probes were indirect in the sense that the machines themselves were macroscopic objects of similar scale to the physicists who built and deployed them. What allowed them to probe the "nano-world" (one-billionth of a meter) where quantum physics would be discovered was a remarkable increase in the precision of their machines and their capacity to manipulate new effects captured in the lab.

A potent example of these new instruments was the spectrograph, a device that can break apart a beam of light and measure its energy content at each component wavelength. Although it had been known since Newton that sunlight could be spread with a prism into the variety of colors that make up the rainbow, the spectrograph allowed the light's intensity (energy) at each color (wavelength) to be measured with high precision. In this way, physicists and astronomers began cataloging the spectra (energy versus wavelength) for everything they could get their hands on. When light from a glowing solid object, like a heated iron bar, was passed through a spectrograph, a smooth curve was found. The spectra showed that energy rose from shorter wavelengths, peaked, and then died off at longer wavelengths. The shape of this curve was found to be universal. Iron, steel, coal, embers of wood: it didn't matter. As long as the heated object was a solid, the spectra always showed the same basic shape, which scientists called a "blackbody" curve. However, when an electric current was passed through a chamber filled with a rarified gas, like hydrogen or oxygen, only a few bright "emission lines" appeared instead of a smooth, continuous blackbody curve. The hot gas emitted no light at any wavelengths other than these few lines. Each chemical element glowed with a unique set of emission lines—its spectroscopic fingerprint.

The first quantum shocks came as physicists tried to use classical mechanics to explain these kinds of data from their spectrographs. It began with the blackbody curve. By the late nineteenth century, physicists had developed a powerful understanding (via James Clerk Maxwell) of light as electromagnetic waves emitted by accelerating charged particles. But all attempts to use this understanding to reproduce the blackbody curve failed. The math behind the physics just gave the wrong result. It was only when Max Planck abandoned a basic assumption of Maxwell's electrodynamics that his new theory recovered the data. Planck had to give up on Maxwell's notion that matter could emit and absorb light waves in a continuous range of energies. Instead, he was forced to assume that vibrating atoms could emit and absorb light only in discrete little bundles or "quanta" of energy. These bundles could not be further divided. For a given frequency of light, there were no half or quarter quanta of energy. Using the hypothesis of discrete energies, Planck's theory beautifully recovered the shape of the blackbody data curve. As a bonus, the idea inspired Einstein to propose, in 1905, that light could be described as both a wave and a particle. Together with Planck's discrete energy bundles, light's wave-particle duality was a direct hit to the neat objectivity of the classical worldview.

Quanta would appear again when Niels Bohr used them in 1913 to explain the emission line spectra from gases. At the time, some physicists were trying to model the atom as a miniature solar system, with electrons orbiting the atomic nucleus like planets orbiting the sun. Unfortunately, when classical electromagnetism was applied to such models, the electrons would radiate away their energy and spiral down toward the nucleus. Thus, classical physics predicted atoms to be unstable. This solar system–like atomic model worked only after Bohr made the radical suggestion that electrons orbited the nucleus on discrete, quantized orbits, like steps on a staircase. A person can't hover ghost-like in between steps any more than an electron can hover between separate orbits. Bohr's model prompted fundamental questions about the world of the very small. What forced electrons to circle their nuclei only at the discrete Bohr orbits? Worse still for the classical views of reality, in Bohr's model, light quanta were emitted (or absorbed) when the electron jumped down (or up) from one orbit to another. How electrons disappeared from one orbit and reappeared in another no one knew. It worked mathematically but didn't make much sense as a physical picture of atomic reality.

Other machines from the workshop showed the quantum to be ubiquitous in the nano-world. An important part of the new technologies was the growing sophistication of what we now call electronics. In the late 1800s and early 1900s, scientists had learned how to make circuits to control electric and magnetic fields with ever higher degrees of precision and accuracy. In this way, although machines were macroscopic devices, they allowed us to manipulate collections of subatomic electrically charged particles with an acuity that transformed them into probes of a world where physics happened at scales billions of times smaller than a human body.

This control led to experiments like those carried out by Otto Stern and Walther Gerlach in the early 1920s, where a beam of silver ions (charged silver atoms) traveled through a specially shaped vertical magnetic field. The electrons in the silver ion were thought to be like electrically charged spinning tops, making them act like tiny magnets.[7] Since magnets interact with magnets, as the electrons passed through the vertical magnetic field, they were deflected by an angle that presumably depended on how fast they were spinning. Classical physics dictates that electrons in the beam would have a continuous range of spin speeds and spin directions. If that were the case, Stern and Gerlach should have observed a smooth range of deflections. Instead, the incoming beam of silver ions split evenly into two halves after passing through the vertical magnetic field. The unavoidable conclusion was shocking: spinning electrons were not at all like classical tops. Their spin was quantized. The data showed that electrons had only two possible spin values: one aligned ("spin up") with the direction of the magnetic field and the other anti-aligned ("spin down") relative to the magnetic field. This is not how the macroscopic world behaves. You can spin a top as fast as you want, and it should be able to point in any direction. Why was the atomic world so different? Classical physics couldn't answer.

What these examples demonstrate is how technologies—machines—provided results that could not be recovered with existing theories. The abstractions developed by classical physics through embodied contact with the world failed to predict or explain the new data these machines provided. In response, physicists developed a new body of theory—a new set of abstractions. As with classical physics, the new quantum mechanics was a triumph. It made predictions that could be verified with astonishing accuracy. Unlike classical mechanics, however, quantum mechanics did not arrive with an easy metaphysics pinned to its cradle. Instead, its mathematical machinery raised fundamental questions that would strike at the heart of the Blind Spot's underlying assumption of a perfectly knowable, perfectly objective, God's- eye view of physical reality and our ability to access it.

The Weirdness That Matters (I): Superposition

Quantum mechanics comes with many different kinds of weirdness, each one violating classical expectations about how the world works. For example, individual quantum events in the nano-world are unpredictable. They just happen without any (known, and possibly knowable) specific cause. Within the current formalism for quantum mechanics, the decay of a radioactive nucleus or the quantum jump of an electron from one orbit down to another are fundamentally probabilistic. Although predictions can be made for the average decay time of a large collection of nuclei, a deterministic prediction of exactly when this or that specific fleck of radium will decay is not in the quantum mechanical cards.

Not all weirdness in quantum physics is equal, however. With regard to the conflict between the metaphysical assumptions of the Blind Spot and the new science of quantum physics, a few key results stand out. The most important ones are the linked questions of quantum superpositions and the measurement problem. To understand their importance, we must briefly introduce the machinery of quantum mechanics, the essential abstractions that drive the theory.

As we have seen earlier in our discussion of classical physics and the arrow of time, the "state" of a system is a key concept in physics. The state represents a complete description of the system in terms of what's needed to predict its future (the future states of the system). Recall that in classical mechanics, the state of a simple system, like an uncharged, unspinning particle, is given by the particle's position and momentum, represented as a point in a six-dimensional phase space. If the particle has a charge, that property gets added to its state as well.

In classical mechanics, the evolution of a system's state is determined by dynamical laws. They are expressed as differential equations, equations that usually describe changes in the behavior of the system in space and time, as it responds to different interactions. For a particle of matter, the equations come from Newton's laws in whatever clothing they happen to be wearing (i.e., the Lagrangian or Hamiltonian formulation). For electromagnetic waves, the dynamical laws come from Maxwell's equations. The importance of dynamical laws is that if we know the state of a system at some initial time, they can be used to predict all future states of the system with no uncertainty.

At least, that's the expectation in principle. In practice (as we explored in the previous chapter), no forward or backward prediction can be absolutely certain, given that we can know the system's coordinates in phase space (typically, its position and momentum) only with finite precision. In particular, for nonlinear systems, small differences in the initial conditions can lead to very large changes in the system's future behavior, as in the famous butterfly effect in climate studies (where the minor perturbation of a butterfly flapping its wings leads to the large-scale difference of a distant tornado several weeks later). Nevertheless, the system is said to be completely deterministic in itself (its prior states fully determine its later ones), despite the limitations on our ability to predict its exact future course.

From the metaphysical perspective of the Blind Spot, what matters most is that the dynamical description of the system's state represents the system's properties in and of themselves. The heart of Blind Spot metaphysics is the existence of a God's-eye view of the system, a perfectly objective ontological view that is independent of our knowledge about the system. The properties of the system, expressed in terms of its state, are real and existent. In other words, even if no measurement is ever made, there is still an independent fact of the matter about the particle's properties (that is, about its state).

Quantum mechanics changed both the description of the state and its relationship to the dynamical laws, collapsing the easy relationship between physics and the Blind Spot and leaving confusion in its wake.

In order to embrace key features of the experimental discoveries of the early twentieth century, including randomness, quantized properties, and the wave-particle duality of matter, the mathematical formulation for the state had to be rebuilt. To see how this was done, consider a simple system with only two possible values of a property (much like what Stern and Gerlach found for the electron spin being "up" or "down"). Following David Albert's description in *Quantum Mechanics and Experience*, we will call this property Color.[8] If a particle (call it *P*) has a color that can be either black or white, then before a measurement is made, the quantum formalism demands that the state of the system *P* be written as

$$|P> = a|\text{White}> + b|\text{Black}>. \tag{1}$$

In words, the state of particle *P* (expressed via the symbol $|P>$) is the number *a* times the state of particle *P* having the color white (expressed via the symbol $|\text{White}>$) plus the number *b* times the state of particle *P* having the color black (expressed via the symbol $|\text{Black}>$). The numbers *a* and *b* are related to the probability that particle *P* is measured to have the color white or black. The equation tells us that before a measurement is made, the particle *P* is in a "superposition" of both black and white colors. If you don't understand what that means, you are in good company. Much of the rest of this chapter turns on exactly that question.

Recall that in classical mechanics, we had both the system's state and a dynamical law that tells us how the state evolves. Quantum mechanics also has a dynamical law, called the Schrödinger equation. This equation gives us two things. First, it gives us a list of all the possible states a measurement could find. In our case, those outcomes are represented by the two states $|\text{Black}>$ and $|\text{White}>$. The Schrödinger equation also tells us about the time evolution of the system—how the coefficients *a* and *b* change (or not) as time advances. Indeed, their values determine the probability that a measurement of the property color will yield white or black (actually, it's the square of their values). In other words, the probabilities for the particle to have the color white (be in state $|\text{White}>$) or have the color black (be in state $|\text{Black}>$) can change in time ($a = a(t)$ and $b = b(t)$).[9] The role of the Schrödinger equation is to specify those changes. As long as no measurement is performed, the particle stays in a superposition whose dynamics (the evolving values of $a(t)$ and $b(t)$) are set by the dynamical law, that is, by the Schrödinger equation.

Now we can begin to explore what it means for the particle to have a color given by equation 1. If we were to ask, "What is the fact of the matter of the particle's color?," then equation 1 is very problematic. It clearly does not mean that the particle's color is black. It also clearly doesn't mean that the particle's color is white. It also doesn't mean that the color is both black and white, or neither black nor white. The truth is that equation 1 doesn't say anything about the fact of the matter of the particle's color. For that, we need to make a measurement. Before we make a measurement, there is no fact.

Unlike classical mechanics, the dynamical law in quantum mechanics is not absolute. It does not hold (or does not seem to hold) for all times and all places. In particular, the one time and place its authority appears to be entirely overruled is when a measurement is made on the quantum system. At that point, the superposed state represented by equation 1 is terminated and a new state obtains. That new state is either

$|P> = |White>$

or

$|P> = |Black>$.

Note that the coefficients a and b are now gone. Probabilities are gone. The system was measured, and a specific result was recorded: either |White> or |Black>. The act of measurement (however we define what that means) has interrupted the Schrödinger equation's hold on the system and forced the superposed state given by equation 1 to "collapse" into one of its two component possibilities.

Although the description given above of the machinery of quantum theory is highly simplified, it captures two essential dilemmas for the Blind Spot. In the first place, the abstractions on which all quantum mechanical calculations are based require that before the system is measured, it lives in the metaphysically problematic superposition state. Superpositions are weird precisely because they are unlike anything in our actual experience of the world. We never encounter objects having multiple and mutually exclusive properties like black and white, up and down, or dead and alive at the same time. Taken at face value, superpositions challenge our idea that things in the world must have well-defined physical properties. That being the case, how can we make sense of the ontological substrate of a world built from superpositions?

Equally challenging is the measurement problem. Physicists love their dynamical laws. Recall that the predictive power of these laws and their ability to recast our messy world in terms of a Platonic realm of pure abstraction lay at the heart of the many triumphs of classical physics. In quantum physics, however, the very act of measuring a system's properties suspends the all-important dynamical law. Why is the link between the state and the dynamical law cut precisely at the point of measurement, seemingly implying the agency of a "measurer"? Quantum physics forces us to reconsider the meaning of both measurement and measurer, and whether there are minimum requirements needed to trigger the collapse of the function describing the system's state (the state function). These are questions completely alien to classical physics.

After more than a century, the nature of superposition and the measurement problem remain profound challenges to the blind spot metaphysics. As a result, various interpretations of quantum physics have been proposed that attempt either to rescue or abandon that metaphysics. In what follows, we explore some of these interpretations relative to the centrality of experience. But before we can take on this task, we must briefly explore one additional weirdness, the linked questions of separability and nonlocality, the bizarre quantum property that an unhappy Einstein famously called "spooky action-at-a-distance."

The Weirdness That Matters (II): Entanglement

State functions can do more than just describe a single particle. They also work for collections of particles, vastly increasing their applicability and their weirdness. Back in 1935, Einstein, along with collaborators Boris Podolsky and Nathan Rosen, noticed something strange about how even the simplest multiparticle state function behaves. Their work, now referred to as EPR, established a new layer of weirdness to the quantum description of reality.[10]

Consider two particles, each having a specific property, which we will call Color and Hardness (borrowing from David Albert's work again[11]).

Color can be either black or white, while Hardness can be either soft or hard. Just as it is possible to create a superposition of a single particle in two states, we can also create a superposition from "two particle states." Here is an example:

$|P_1P_2> = a|\text{White}_1>|\text{Hard}_2> + b|\text{Black}_1>|\text{Soft}_2>.$

In words, the state of particles P_1 and P_2 (expressed via the symbol $|P_1P_2>$) is the number a times the state of particle P_1 having the color value white and P_2 having the hardness value hard (expressed via the symbol $|\text{White}_1>|\text{Hard}_2>$) plus the number b times the state of particle P_1 having the color black and P_2 having the hardness value soft (expressed via the symbol $|\text{Black}_1>|\text{Soft}_2>$). The numbers a and b are, once again, related to the probability that the combined state of particles P_1 and P_2 is either $|\text{White}_1>|\text{Hard}_2>$ or $|\text{Black}_1>|\text{Soft}_2>$.

At first, this combined state may look like a more complex version of equation 1. On closer inspection, however, it expresses something bizarre about the possible behavior of the two particles. Imagine we make a color measurement on particle P_1 and find it is white. That means after the measurement, the new state must be

$|P_1P_2> = |\text{White}_1>|\text{Hard}_2>.$

Notice how the collapse of the state function in this case means that particle P_2 must be in a hardness state with value Hard. We can know this even without making a measurement on particle P_2. Knowledge about P_2 is based on knowing the state of the wave function before anyone messed with particle P_1. Likewise, if we had measured particle P_1 to be black, the state afterward must be

$|P_1P_2> = |\text{Black}_1>|\text{Soft}_2>.$

Once again, making a measurement on particle P_1 forces particle P_2 to take a specific value. So what is the problem here? As Oliver Morsch points out, we never said anything about where the two particles are positioned in space.[12] Particle P_2 could be on the other side of the galaxy when we make our measurement of particle P_1. Despite that distance, the instant the measurement is made showing particle P_1 to be white, particle P_2 must instantly take the property of being hard. The two-particle quantum systems described above are said to be "entangled." The particles and their properties are not separable. Somehow, they form a whole whose evolution is intimately coupled even when they are separated by distances far greater than light could have traveled during the time it takes to make a measurement.

This "spooky action at a distance" deeply bothered Einstein. Given what he had done with the theory of relativity, making gravity a local phenomenon related to the curvature of space-time, he was sure that nature had to be local, meaning that no object and no information could travel faster than light. That's why he, Podolsky, and Rosen thought that two-particle state functions, as in our example, could be used to show that quantum mechanics was incomplete. They argued that there must be some "hidden variables" lurking below the usual formulation of quantum mechanics that explained how widely separated particles could behave like they were connected, a kind of invisible conductor telling the particles what to do.

In the early 1960s, Irish physicist John Bell found a formula demonstrating how to distinguish quantum mechanics and its nonlocal entanglement from EPR's belief that a classical, local, hidden-variable approach must lie beneath quantum physics (in contrast to nonlocal hidden variables, as discussed below). While it took almost two decades for Bell's formula to be tested in the laboratory, the results then, and now, came down conclusively on the side of nonlocal quantum entanglement. Reality is clearly weird.

Toward Interpretation

There were no long-running battles over the interpretation of classical mechanics. Although philosophers certainly found reasons to philosophize over details, for physicists themselves and the interested general public, its interpretation seemed clear. The world was composed of tiny bits of matter that moved through space-time according to the laws of mathematical physics. End of story.

Quantum mechanics was different. The atoms of quantum theory bear no metaphysical resemblance to those of the Greeks. While the formalism of quantum mechanics provides the most stunningly accurate answers science has ever yielded, it's impossible to draw a simple picture of what that formalism represents. What does the state function refer to? What is a superposition in terms of real things in the world? What does entanglement mean for those real things in the world? Most of all, why does measurement matter?

Unlike its classical counterpart, quantum mechanics does not come with a ready interpretation stuck to its side like a game of metaphysical pin the tail on the donkey. As a result, for more than a hundred years, physicists and philosophers have proposed a long string of conflicting interpretations. Each interpretation attempts to account for the meaning of superpositions, entanglement, and the measurement problem. In the rest of this chapter, we review the most popular of these interpretations in light of the Blind Spot. Our purpose is not to advocate for one interpretation over another. Instead, we want to understand how these interpretations respond to the challenge posed by quantum mechanics by focusing on what they privilege. Behind each interpretation stands a worldview, a viewpoint about what matters in science and why.

Most quantum interpretations fit into one of two categories regarding what they privilege. The categories turn on the meaning of the state function, which is often represented by the Greek letter Ψ (Psi). Interpretations that want the state function to be a real "thing in-itself," existing out there in the world, are waggishly referred to as "psi-ontological." Opposing these are interpretations that see the state function as a measure of our knowledge of the world. These are called "psi-epistemological." Ontology versus epistemology: this battlefront in the hundred-year-old quantum interpretation wars has never known resolution.

Before we discuss how the conflicting interpretations relate to the Blind Spot, it's worth noting that the scientific deployment of quantum mechanics does not require an interpretation. It may seem strange, but it's entirely possible to ignore questions about what superpositions, entanglements, and measurement mean and just use the formalism to design and analyze experiments. This "shut up and calculate" approach, so-named by physicist David Mermin, works just fine.[13] In fact, for many years, concern about the "foundations of quantum mechanics" (how to interpret its deeper meaning) was seen as a career killer for young physicists. Over the past few decades, however, quantum information theory and the tantalizing possibilities of quantum computing have become leading research fields. In these domains, the preservation and manipulation of individual superpositions (known as q-bits) is a central concern. As happened during the nineteenth century, when technological applications linked to the Industrial Revolution brought forth thermodynamics, entropy, and the arrow of time, the new quantum technologies are bringing foundational questions, which were mostly avoided a generation ago, back to the forefront.

Psi-Ontology (I): Random Collapse and GRW Interpretations

One way to understand the measurement problem is to focus on the abrupt stoppage of the dynamical law that comes when a measurement is made. Is there a way to understand the collapse of a superposed state function that does not demand the abandonment of Schrödinger's equation? In other words, for those who favor a psi-ontological viewpoint, the effort is aimed at preserving ontology, not explaining measurement. The goal is to find an interpretation of quantum mechanics that ensures that things in the world, including the wave function, have properties in themselves. One attempt follows the so-called dynamical reduction program, also known as spontaneous collapse theory, which seeks a purely ontological dynamical mechanism forcing the superposition of many states into a single value— the state observed that may be recorded by a measurement but is in no way dependent on it.

The most popular approach along these lines was developed by Giancarlo Ghirardi, Alberto Rimini, and Tullio Weber in a 1986 paper.[14] Their theory, called the GRW theory, adds new terms to the Schrödinger equation. These new terms model interactions with the environment (the measuring device) that spontaneously drive superposed quantum systems into a collapsed state. This kind of "wave packet reduction" is governed by parameters in the new terms, which, in principle at least, would take on the status of new constants of nature (akin to Newton's gravitational constant G or the speed of light c).

One of the most important features of GRW is that the new terms added to the dynamical law are stochastic. This means that each interaction between the superposed quantum system and one of the vast number of atoms in the measurement device holds the possibility of collapsing (that is, reducing) the superposition. This approach has merit, because in principle, a single interaction between any atom and the superposed quantum system can trigger the collapse. As the number of interactions per time increases, the probability that one of them triggers the state collapse approaches absolute certainty. This feature of GRW also explains the transition in behavior from the long-lived superpositions in the nano-domain to the macro-domain of everyday experience, where systems are never observed in superpositions. Isolated nano-systems made of a few atoms can be maintained in superpositions, coupled macro-systems made of zillions of atoms cannot.

The technical problem faced in GRW models is to provide a full mathematical account of how, where, and when interactions between the quantum system and the environment trigger modifications in the quantum system's state function. Since the original GRW paper, different approaches have been suggested with varying degrees of success. Regardless of the details, the important point is that GRW models constitute an alternative version of quantum mechanics that features modifications of Schrödinger's original dynamical law.

From the perspective of Blind Spot metaphysics, particularly its commitment to the idea that fundamental physics provides access to reality apart from any human perspective, the most noticeable feature of most versions of GRW is that the superpositions themselves are never accounted for. The basic question of what it means for a system to be in a state without definite physical properties remains unanswered. Instead, superpositions are simply claimed to exist for short enough times relative to macroscopic objects like us: there is no attempt to elucidate the meaning of quantum superpositions. Although the added terms and constants allow for an ontic view of the wave function, the basic ontological concern is not addressed. More generally, from the perspective of our exposition of the Bind Spot, GRW, and, in fact, all psi-ontological interpretations, clearly make use of a surreptitious substitution, the substitution of a mathematical quantity developed to describe physical data—the wave function—for a real thing in the world, without any empirical evidence that this substitution is justified. In doing so, it elevates hypothetical parameters that control the wave function collapse to (unobserved) new constants of nature on a par with, say, the speed of light.

Psi-Ontology (II): Pilot Wave Theories

In the 1920s, Louis de Broglie proposed another way to address the indeterminacy of quantum mechanics that was rediscovered and further refined by David Bohm thirty years later.[15] The de Broglie–Bohm theory, also known as Bohmian mechanics, calls for the addition of a "hidden variable" to restore determinism to quantum mechanics. Like Einstein, Podolsky, and Rosen in 1935, de Broglie and Bohm considered the original quantum formalism to be incomplete. They called for other, hidden aspects of the quantum dynamics that can provide a complete description of atomic phenomena without any reference to measurements and observers.

The de Broglie–Bohm theory begins with the famous wave-particle duality of quantum theory. The Schrödinger equation literally describes waves propagating through space, represented by the wave function Ψ. Unlike waves in water or waves of electromagnetic fields, however, what emerges from Schrödinger's equation are waves of probability for a system to attain properties with this or that value (like a particle's position). This is what the coefficients a and b in the state function shown in equation 1 refer to. It was Max Born who first saw that the square of these coefficients (a^2 and b^2) yielded probabilities that different physical properties of the state function would be found in a measurement (for example, |White> or |Black> in equation 1). What the de Broglie–Bohm theory adds to this picture is a new equation—a new law and a new entity—for the description of quantum systems. For the de Broglie–Bohm theory, particles are real entities, and, in a sense, what the original formalism of Schrödinger provides is a kind of "pilot wave" that guides the motion of the particles.

This pilot-wave view was formalized in Bohmian mechanics. Physical properties of quantum particles like position and momentum are deterministic, their values guided by the pilot wave. With this addition, superpositions no longer need to be considered real things in the world. Particles are really little bits of matter as the Greeks imagined, with their motions guided by the pilot waves, like an orchestra conductor guiding musicians. In this way, dynamics becomes complete, meaning that there is always an answer to what properties a particle has before a measurement is made.

For many years the de Broglie–Bohm theory was forgotten or ignored due to a number of misconceptions about its meaning and validity. Recently, however, a growing understanding that the theory can provide an alternative formalism to quantum mechanics has reenergized research on its viability.

One question about Bohmian mechanics concerns locality—that interactions between widely separated objects cannot occur faster than the time a light signal crosses the distance between them. Locality is deeply enmeshed with the notions of causality and the direction of time, as we saw in the previous chapter. Since no physical cause can travel faster than light, most physicists demand that accounts of reality must always be local. We have seen that Bell's theorem forbids any hidden variable theory to be local, a result now confirmed by multiple experiments. Consequently, Bohmian mechanics is nonlocal. This means that the whole universe somehow participates in determining the behavior of every single material particle. Although Bell was an avid proponent of the deBroglie–Bohm theory, many physicists find the need for a nonlocal dynamics to detract from its value as an ontological interpretation of quantum theory. For such physicists, the price seems too high.

Another point of contention for the de Broglie–Bohm theory is the ontology of pilot waves and particles. These waves are themselves hyperdimensional entities with no obvious relationship to other things that physics takes to be real. After a measurement, for example, the de Broglie–Bohm mechanics retains empty or "ghost branches" of the pilot wave describing possible experimental outcomes that are not observed—the "other" states in the superposition. These branches of the state function retain their ontological status, still floating around out there, real but unrealized.

Psi-Ontology (III): Parallel Realities and the Many-Worlds Interpretation

First articulated by Hugh Everett in 1957, the many-worlds interpretation (MWI) proposes a straightforward explanation of the collapse of a superposed state function: it never happens. The goal of MWI, like that of other psi-ontological interpretations, is to preserve ontology. Thus, rather than have the dynamical law represented by Schrödinger's equation superseded by a measurement carried out by an observer, the universe itself "splits" into parallel branches at the moment of measurement. Each branch represents a world in which the observer records a different measurement outcome, with each outcome identified with a different piece of the original state function. Thus, a measurement of the quantum system represented by equation 1 would create a branch (a world) in which observers would record the apparatus reading "|White>" and another in which observers see the apparatus reading "|Black>." From that point on, each parallel branch continues to evolve, only to branch again as new quantum measurements are performed.

What makes the MWI attractive to some physicists and philosophers is that it preserves the ontological status—the objective reality—of both the state function and the dynamical law that determines its evolution.

One can talk about a "state function of the universe" that represents a complete description of all particles and processes. In the MWI, this state function is determined only by initial conditions (given at the Big Bang), with subsequent evolution given by a souped-up Schrödinger's equation (the relativistic, field-theoretic version of the dynamical laws). Nevertheless, in the absence of an understanding of the quantum nature of gravity, it is not clear what a wave function of the universe means. Another quality that makes MWI attractive to its adherents (and unattractive to its critics) is that the indeterminacy associated with quantum interactions also disappears. By assuming the state function and its dynamical law to have absolute authority over the evolution of the universe, quantum mechanics becomes fully deterministic. The question about how one accommodates this determinism to the probabilistic nature of quantum physics remains a topic of discussion.

The "joining" of a quantum system and a measuring apparatus creates the parallel branches in the MWI. Interactions generate subsystems comprising the apparatus and the quantum system in different states. It is this superposed state of the subsystem that evolves into multiple parallel (i.e., noninteracting) branches. A key to this splitting process is decoherence, in which an isolated, superposed quantum system interacts with the "bath" atoms in the much larger, macrosized measuring apparatus. Decoherence plays an essential role as the physical mechanism causing the branching of an initial entangled state of a quantum system and the measurement apparatus into coexisting parallel states in noncommunicating universes, each representing a possible measurement value. In this way, the many worlds of MWI are not parallel universes but parts of a single universe that are invisible to each other. The other worlds become like ghosts to each other, present but unable to interact.

Given its apparent other-worldly nature (pun intended), the MWI has been criticized on a number of fronts. The most obvious criticism invokes Occam's razor, which warns against multiplying unneeded entities in an explanation. Defenders of the MWI, however, argue the opposite: that it is the nonlocal collapse of the wave function during measurement that is an unnecessary added entity to quantum mechanics. The price the MWI pays to save the ontology of the state function and a deterministic worldview is the creation of a near-infinite number of new worlds, each one containing a copy of you (and every other person), with its own unique consciousness and believing it is you. This ontology is a price too high to pay for many physicists and philosophers. There are also technical difficulties that arise in the treatment of probabilities in the MWI. In particular, it is not clear how to assign relative probabilities to the different branching worlds that reflect the results of experimental measurements. From our perspective, the elaborate ontology of MWI is symptomatic of the Blind Spot and the surreptitious substitution. The commitment to maintain the ontological primacy of the wave function forces MWI adherents to make a choice that leaves them with a universe full of ghosts.

Between Epistemology and Ontology: Relational Quantum Mechanics

A recent addition to the menu of interpretations worth noting in our context is relational quantum mechanics (RQM), introduced by Carlo Rovelli in 1996.[16] RQM is somewhat orthogonal to both the psi-ontological interpretations we have just explored and the psi-epistemological interpretations we cover in the following sections.

Like psi-epistemological models, RQM does not take the state function to represent a new entity in the universe. It is a tool for computation that has no intrinsic ontological status. At the same time, like classical mechanics, RQM holds that a physical system has intrinsic properties like position and momentum existing "out there" in the universe. RQM's innovation is to claim that such properties exist only when interactions occur with other systems. Moreover, these properties have meaning only relative to those other systems with which the interaction occurs. Thus, the properties of physical entities are always relational properties.

RQM does not require that any special status be given to observers or measurement. Any physical system can act as an observer because measurements are now simply interactions. The price to be paid for this view is that RQM is not strongly realistic in the metaphysical sense. Physical systems take on real properties only at the moment of an interaction (a flash). In between these interactions, nothing can be said about the real value of the system's properties. The relational nature of the interpretation also means that "different observers can give different accounts of the same sequence of events."[17]

RQM has elements in common with many of the other interpretations we are discussing here. Like the many-worlds interpretation, RQM "indexes" the branches of the superposition, but rather than create new worlds with each measurement, this enumeration serves to identify the network of interactions and the flashes of relational properties produced. Like the Copenhagen interpretation, which we cover next, it focuses on the interaction between an observer and a quantum system. It "democratizes" the interaction, however, by not assuming a special role for a classical macroscopic world and by making any interaction play the role of an observer. Finally, like QBism, the last interpretation we cover, RQM focuses on the informational nature of the state function and insists on dropping questions that are meaningless given the quantum formalism. Unlike QBism, however, it does not emphasize a role for agents or observers.

Psi-Epistemology (I): The Copenhagen Interpretation and Quantum Orthodoxy

"Orthodox quantum mechanics" is often used to identify the standard view, also called the Copenhagen interpretation, in contrast to the various psi-ontological views we just reviewed. The Copenhagen interpretation has been the operational way of thinking about quantum mechanics for more than eighty years. The ideas behind it appear in many textbooks, even if that term is not used.

In truth, there is no single "Copenhagen interpretation." Instead, the term refers to an amorphous collection of ideas and perspectives associated with some of the founders of quantum mechanics, including, in particular, Werner Heisenberg, Max Born, Wolfgang Pauli, Eugene Wigner, and, most of all, Niels Bohr, who was based in Copenhagen.[18] There were significant differences among these thinkers about the nature of the state function, the role of measurement, and the distinction between the classical and quantum worlds.

Niels Bohr was notoriously elliptical in his writing style, making it difficult sometimes to pin down his arguments on key issues concerning the meaning of the quantum formalism. Bohr thought atoms were real, meaning they were not "heuristic or logical constructions."[19] But in the atomic and subatomic domains of quantum mechanics, the concepts of classical physics no longer refer to objects in and of themselves. Everything must be referred back to the experimental arrangement. As a consequence, all of our knowledge about quantum systems was context dependent. The experiment, set up by the experimentalist, could not be taken out of the picture. This perspective draws an important distinction between the classical and the quantum realms, where the system under study and the device used to carry out that study could be cleanly separated. Thus, for Bohr and the Copenhagen interpretation, there is a cut—a separation between the classical and quantum realms of phenomena. As one progresses from quantum scales to those associated with the macroworld, the descriptions of quantum mechanics must be complementary to (meaning, able to recover) those of well-tested and reliable classical mechanics.

In the Copenhagen interpretation, the state function Ψ does not allow for a picture of what is occurring on nanoscales: the state function does not represent a thing or an entity. Instead, it is symbolic of our understanding of the context-dependent experimental arrangement, being by nature epistemic. It is an expression of our knowledge of the quantum system, gathered through the experiments we perform. Note that such knowledge is objective, in the sense that two experimentalists can perform the same experiments and compare their results. If the experiments yield the same answers, then those answers constitute objective knowledge.

In relation to the measurement problem, the basic idea of the collapse of the wave function comes from the Copenhagen interpretation. This is what we mean by referring to it as an "orthodoxy." In 1932, John von Neumann offered what has become a standard account of collapse by distinguishing between type 1 processes—those associated with measurement—and type 2 processes—those associated with the evolution of a quantum system via the Schrödinger equation.[20] He further suggested that one should distinguish the quantum system, the measurement device, and the actual observer. Where the dividing line between these is drawn was contextual, he suggested.

Eugene Wigner later picked up on the ambiguity concerning the apparent role of an actual observer (a mind) in his famous "friend paradox."[21] An observer records a measurement of a system that was in a superposition, thereby collapsing the state function. The observer's friend watches the lab in which the observer makes the measurement. Before the friend looks and learns the result of the measurement, however, she would describe the observer and the quantum system in a superposed state. For Wigner, this indicated some role of consciousness in the collapse process.

This view, however, was never Bohr's position. Although Bohr saw the state function fundamentally as a context-dependent description of quantum phenomena, he was not likely to have seen any role for nonphysical mental activity directly affecting quantum systems (or any physical system).

Thus, the Copenhagen interpretation is actually a mix of sometimes divergent views among some of the founders of quantum mechanics. What they share, however, is a definite sense that the state function belongs to our knowledge rather than to reality-in-itself apart from the act of knowing: it represents what we can say about the world through our measurements, not how the world is in itself. Most of the viewpoints under the Copenhagen interpretation banner recognize that quantum mechanics profoundly changes how we should understand the relationship between the system under study and those who are studying it. In this way, since its inception, quantum mechanics has led some of its most creative theorists to believe that the new science demands going beyond the metaphysics of the Blind Spot.

Psi-Epistemology (II): Qbits and the Primacy of the Subject

Quantum Bayesianism, or as it is now called, QBism, is a relative newcomer to the interpretation contest.[22] The original use of "Bayesianism" in its name came from its emphasis on taking the probabilities appearing in quantum mechanics at face value and interpreting them in a Bayesian framework. (The term *Bayesian* comes from the eighteenth-century mathematician and theologian Thomas Bayes.) Probabilities represent a state of knowledge or belief for Bayesians instead of a frequency or propensity of some phenomenon. Although there are various streams of Bayesian thinking concerning what a probability distribution means (which is one reason that quantum Bayesianism was renamed), the one that matters for QBism is that probabilities reflect the current state of knowledge of an agent about the world, as they make bets about what will happen in the future. As physicist E. T. Jaynes put it in describing the Bayesian view of probability for science, "Probability is a theoretical construct, on the epistemological level, which we assign in order to represent a state of knowledge, or that we calculate from other probabilities according to the rules of probability theory."[23]

From the perspective of QBism, much of the spooky nonlocal weirdness of quantum mechanics can be resolved by accepting that quantum mechanics (and all scientific endeavor) ultimately refers back to the experiences of the agents using the theory. Thus, the quantum state is no longer a property of the world in and of itself. Instead, these states must be associated with the agents using quantum mechanics to do experiments concerning the microphysical world. Each agent will assign quantum states to a system depending on the context in which that agent finds herself. Quantum states are not elements of reality but represent the degrees of belief an agent has about the possible outcomes of their making a measurement. Hence, the ontological weirdness of superpositions disappears because the state function is not ontological. The question, "What is it like to be in a superposition?" doesn't arise because superpositions aren't states of the world. They are states of knowledge (or, more precisely, states of belief, since Bayesians generally understand probabilities to be degrees of confidence or "credences" that agents have in propositions given the evidence available to them). QBists thus claim that entanglement loses its paradoxical nature. What is learned from the measurements made by observers on pairs of particles that have interacted and then separated is how quantum probabilities reveal the structure agents must deal with in making their choices about the world. Understanding that structure is what QBists claim to be the ultimate goal of their project.

QBists would also claim that measurement is no longer a problem in their framework because measurements are simply acts an agent performs on the external world. This makes "each measurement outcome . . . a personally experienced event specific to the agent who incites it."[24] Thus, there is no longer a need to try to identify the "shifty split" that marks the cut separating quantum system, measuring apparatus, and the environment.

QBism takes what appears to some to be a radical subjectivist approach not only to the quantum formalism but also to science itself. Given this emphasis, one criticism claims that QBism is solipsistic. QBists reject this by arguing that the external world is a central postulate of their framework. In the QBist view, quantum theory is always about *"an agent's interactions with the outside world*; the formalism of quantum theory makes no sense otherwise."[25] As physicist N. David Mermin puts it, "QBism puts the scientist back into science."[26]

To state that your understanding of the world rests on your experience is not to say that your world exists only within your head, as some articles about QBism have wrongly asserted. You construct your picture of the world from many ingredients, including the impact of the world on your experience and how and when the world responds to the actions that you take. When you act on your world, you generally have no control over how it acts back on you.

Although you have no direct personal access to someone else's experience, an important component of your private experience is the impact your efforts to communicate it have on others. Science is a collaborative human effort to find, through our individual actions on the world and our verbal communications with each other, a model for what is common to all of our personally constructed worlds. Conversations, conferences, research papers, and books are an essential part of the scientific process. Thus, QBism is about agents living and acting in a shared world, and that world always pushes back. There is nothing solipsistic about QBism.

QBism is also sometimes claimed to be nothing but an extension of the Copenhagen interpretation. QBists respond to this by first noting that there is no one Copenhagen interpretation. More important, however, unlike any version of the Copenhagen interpretation, QBists are explicit about introducing agents or "users" of quantum mechanics and making them central to an understanding of the quantum formalism. In Mermin's words, "Since every user is different, dividing the world differently into external and internal, every application of quantum mechanics to the world must ultimately refer, if only implicitly, to a particular user. But every version of Copenhagen takes a view of the world that makes no reference to the particular user who is trying to make sense of the world."[27]

Uncertain for a Century: The Quantum and the Blind Spot

One of the most remarkable aspects of quantum mechanics, and its opulent basket of weirdness, is how resilient it is against the interpretation controversies. After more than a century, we are still uncertain about how to interpret the remarkable success of the mathematical formalism of quantum physics. Given one hundred years of argumentation, however, what is clear is the steep price that must be paid for any attempt to interpret that formalism. That price comes at the expense of the Blind Spot.

Each interpretation of the quantum formalism forces adherents to take a giant step away from the straightforward objectivism of the Blind Spot and its grounding in classical mechanics. For example, the price the many-worlds interpretation pays for making the state function a real thing in the real world is an unaccountable proliferation of worlds with innumerable copies of every one of us. The price QBism pays for unraveling the weirdness of superpositions and measurement is an acknowledgment that agents must be central to any description of physics. In both cases, and for very different reasons, the Blind Spot's vision of classical physics with just one perfectly knowable universe must be abandoned in favor of a new set of assumptions about the nature of the world and our relationship to it.

Quantum mechanics forces us to ask a difficult metaphysical question: What price is too steep? Which assumptions about the world are too important to give up? Better yet, which new principles are now so important that they must be woven into the fabric of physics? From the perspective we've been developing in this book, it is clear that any drive to reify the abstractions of mathematical physics will lead to problems. Whitehead made this point even before the quantum interpretation controversies really got going. His "fallacy of misplaced concreteness" was simply a recognition that the substitution of mathematical abstraction for concrete experience should be expected to lead to puzzlement and paradox because it is an unnatural bifurcation of nature. In quantum mechanics, we see this fallacy played out as the reification of mathematical constructions forces some interpretations into what seems like bizarre territory. One must ask why some scientists, who are usually keen to stay close to data and experiment in other domains, are willing to stretch so far beyond experience when confronted with quantum mechanics. Why is something like the many-worlds interpretation, which seems ontologically extravagant and holds no obvious experimental advantage, chosen as an interpretive framework by many physicists and philosophers? Why is that price, an explosive proliferation of parallel worlds, the one they are willing to pay?

The answer is the Blind Spot. As we have seen, the rise of classical physics entailed a selective forgetting of the role of experience as the ground of all scientific practice. The ascending cycles of abstraction from Newton to Poincaré meant experience was reduced to mere observation or the recording of data. Objectivist ontology became king as scientists grew accustomed to assuming that the creations of their mathematical physics could be treated as timeless laws held in the "mind of God" and viewable from a perfectly objective, perfectly perspectiveless perspective—a "view from nowhere."[28] Thus, when quantum mechanics appeared from the same experimental workshop that had created the triumph of classical physics, many scientists believed their job was to defend the ontological heights and equate reality with the abstract formalism.

The recognition that the Blind Spot has been a foundational mistake running through the culture of science (as opposed to the practice of science itself) since its modern inception throws a particular light on the quantum mechanics debates. When we understand that experience had been forgotten but could remain forgotten only for so long, we should expect that interpretations returning experience to its proper centrality in science are the logical response to the challenge quantum mechanics poses. Seen in this light, the epistemological price that QBism imposes seems to be a good wager rather than a forbidding one. It not only recognizes the centrality of experience but also provides a framework for understanding how new physics might emerge from that recognition.[29]

Although we are not claiming that QBism is the correct interpretation of quantum mechanics, we are claiming that it is a move in the right direction. Together with elements of RQM, it addresses the fundamental question posed by quantum mechanics: the relationship between properties of the world and our experiences of the world. We can thus begin to see, slowly emerging within the long-standing debates about quantum physics, an example of how the Blind Spot has forced certain limited ways of thinking about science to be accepted and how those ways may now perhaps be overcome.

5 Cosmology

Returning to Time

We now turn from the small to the large, from the quantum to the cosmos. Focusing on cosmology brings us back to the topic of time and the issues concerning lived time versus clock time. Earlier we explored time in classical and statistical mechanics. We saw how powerful abstractions allowed physicists to make extraordinary progress in capturing phenomena related to celestial mechanics and the exchange of heat between bodies. Hidden within these successes, however, were potent elements of the Blind Spot, particularly when scientists tried to deal with vexing questions about the direction of time. Here we continue the story into the modern era with the rise of relativistic theories of gravity and the cosmological science it made possible. We begin with Albert Einstein and the scientific revolution his theory of relativity brought about in our understanding of time and the cosmos.

Einstein and the Localization of Time

All models fail. This is to be anticipated, given that models are idealizations of reality. When pushed hard, they eventually show their cracks. And that's a good thing. New science emerges from such cracks, sometimes even whole new worldviews. Physicists are trained to expect and even welcome such fragility in their models and theories, although in reality, things are never so simple. Being human, physicists get attached to their favorite models, and emotions often run high when the viability of a beloved model is threatened. This is why the scientific enterprise is designed to recognize its own weaknesses: the final decision concerning the viability of a model or a theory is empirical and community based, not personality based. Sooner or later, as new data are gathered and analyzed, egos are pushed aside and consensus prevails. At any rate, this is the way things are supposed to work, and for the most part they do.

What is more difficult, however, is for scientists to see below the foundations of the scientific workshop where these debates are carried out. Basic assumptions about the measurer and the measured, the map and the mapmaker, go unchallenged because they have become invisible. They have fallen into the Blind Spot. This forgetting—the amnesia of experience—was also present in the debates over time and relativity as the cracks in the Newtonian worldview appeared and Einstein's revolutionary work emerged and took hold in science.

Three cracks in the Newtonian worldview of mechanics and gravitation are well known: first, fast motions call for corrections from the special theory of relativity; second, strong gravity calls for corrections from the general theory of relativity; and third, objects of molecular, atomic, and subatomic dimensions call for an alternative mechanics based on the rules of quantum physics. The Newtonian worldview of absolute time and space, so useful at describing the world of our everyday lives, is, in effect, a blurred mapping of reality. It works well because relativistic and quantum corrections become sizable only at extreme circumstances far removed from our perception of the world. The cracks in the foundations of the Newtonian framework are well hidden. Once they were exposed, however, all three cracks called for a deep revision of fundamental concepts, developed during the first three decades of the twentieth century. Time, as we will describe in this chapter, gained a plasticity that undermined Newton's notion of absolute time. Time became local and frame dependent, each observer's clock ticking at its own rate.

Einstein showed that Newton's absolute time, a universal clock ticking inexorably at the same rate for all observers, in all regions of the universe, is an approximation, valid only for slow relative speeds and weak gravity, which happen to be the conditions in which we live our lives. Still, the theories of relativity and quantum physics are necessarily based on experience, particularly on experiences created using the tools of the scientific workshop.

Einstein's monumental achievement was to create a conceptual framework that allowed different observers to compare their clock times, including when moving at dramatic speeds and in strong gravitational fields. Nevertheless—and this is essential to our argument—the time of relativity and quantum theory retained the mathematical features of Newton's abstract, physical time: time is represented on a continuous time line and built out of instants with no duration. The equations of special and general relativity still assume implicit causality and necessitate initial conditions in order to be solved. Time retains its order, but it can still flow in either direction, past to future or future to past. With Einstein, Newton's single, universal time became an approximation, as physical time flows differently according to local conditions. The stronger gravity is, the slower time passes. We can interpret both the special and general theories of relativity as translation devices that allow observers moving with different speeds or in different gravitational fields to compare their local measurements of time so that the laws of physics remain universally valid. Understanding the inevitability of local variations in measurements of time and space, Einstein restored the integrity of a universal physical narrative, using the constancy of the speed of light as its cornerstone.

No wonder Bergson and Einstein clashed on the nature of time in their famous public debate in 1922.[1] Bergson pressed Einstein on what we are calling the Blind Spot, the neglect of experience in the description of natural phenomena, in this case in the conceptualization of time. Einstein shrugged away Bergson's critique as being psychological, and thus essentially useless to physics.

Einstein is generally credited with having prevailed in the debate, while Bergson's reputation suffered and his influence waned.[2] Bergson is said to have failed to understand relativity theory and its mathematical formulations in his book *Duration and Simultaneity*.[3] Bergson's analysis may have contained some errors, but careful examination of the book and the debate with Einstein does not bear out this lopsided judgment.[4] As physicist C. S. Unnikrishnan writes, "His rigour as a mathematically competent philosopher addressing questions in a physical theory was stringent, and he was meticulous in his analysis."[5] The debate's philosophical and scientific import deserves further scrutiny. From our perspective, a hundred years later, the debate represents a missed opportunity for moving our scientific worldview beyond the Blind Spot. Indeed, the debate brings to light how the advance of post-Newtonian science, represented at the time by Einstein, already demanded supplanting the Blind Spot. Bergson understood this, despite his mistakes. Bergson was right, even if he was also wrong, and Einstein was wrong, even if he was also right. For these reasons, we need to take a closer look at what was really going on in the debate.

Simultaneity in the City of Lights

April 6, 1922. This was the momentous date on which Einstein addressed the Philosophical Society of Paris. Bergson, an intellectual celebrity known for his impressive oratory, said he came only to listen and not to speak.[6] According to Maurice Merleau-Ponty, however, "The discussion flagged" and Bergson was called on to address the gathering.[7] Yielding to the "amicable insistence" of the society, Bergson rose and presented extemporaneously a few ideas from his soon-to-be published book *Duration and Simultaneity*, which was devoted to a "confrontation" between his concept of duration and Einstein's views on time.[8]

Bergson began his remarks by declaring his admiration for Einstein's work: "It appears to me to demand the attention of philosophers as much as of scientists. I see in it not only a new physics, but also, in some ways, a new manner of thinking."[9] He made clear, at the end of his speech, that he had no objection to Einstein's definition of simultaneity or to the theory of relativity. Rather, his concern was to establish that "once the theory of relativity is admitted as a physical theory, all is not finished. We still have to determine the philosophical significance of the concepts it introduces."[10]

A hard-nosed physicist might wonder why we have to do this. Why isn't it enough to say that the concept of time designates whatever a clock measures, and leave it at that?

That was more or less Einstein's response. Bergson, in his impromptu remarks, had proceeded to outline the idea of a universal time of duration and had argued that the relativistic definition of simultaneity actually implied the intuitive concept of absolute simultaneity as given in the experience of duration: "absolute in the sense that it does not depend on any mathematical convention, on any physical operation such as the setting of a clock."[11] Einstein listened carefully, but his reply was perfunctory. He briefly summarized his understanding of what Bergson had said and then stated, "There is no time of the philosopher; there is only a psychological time different from the time of the physicist."[12] By "psychological time," Einstein meant the subjective perception of time's passage, dependent on individual factors such as how our perceptual systems respond to sensory stimuli and to our emotions or mood.

Bergson must have been chagrined. Einstein hadn't responded directly to his arguments but instead had asserted that the experience of duration is just a psychological phenomenon. Bergson's concern was the temporality of time, time's passage. Einstein refused to countenance it other than as a mere psychological phenomenon. There are perceived events, which Einstein declared to be "mental constructions," and there are the objective events of physics measured by clocks.

But Einstein hadn't addressed Bergson's point, let alone his arguments. Bergson was saying that we can't make sense of the concept of clock time without referring to something that is not a reading of a clock, namely, "the moment in which we find ourselves, the event that is occurring," which we know only as duration.[13] That is the moment in which any clock must be read because clocks don't read themselves. Bergson was pointing out that you must experience time as duration before you can abstract time as a measured point in the abstraction of a space-time diagram. Our experience of time precedes the mathematization of time needed to build a clock.

Some physicists and philosophers continue to deny this point, claiming that the observer in relativity theory doesn't have to be a living being or a conscious subject, but could "just as likely . . . be a photographic plate or a clock," or a computer.[14] But this is a grave mistake, one that perfectly exemplifies the amnesia of experience in the Blind Spot. Nature does not produce measuring devices except via human beings and their intentions. Unless what the photographic plate, clock, or computer registers is itself registered in experience, it has no scientific meaning. It has no value in the scientific workshop.

Bergson's point was that it's only within the duration of the present moment—the experienced now—that we establish the simultaneity between an event we perceive and a clock we read. In his words:

> The simultaneity between the event and the clock's reading is given by the perception that unites them in an indivisible act. It essentially consists of the fact— independent of the setting of the clocks—that this act is *one* or *two*, depending on our will [that is, we can take our momentary perception comprising the event and the clock reading either as a whole (hence one), or as divided between the event and the clock-reading (hence two), but without the perception's durational unity being split apart.] If this simultaneity did not exist, the clocks would serve

no purpose. We would not manufacture them, or at least no one would buy them. For one only buys a clock in order to know what time it is; and "knowing what time it is" consists in observing a correspondence, not between one clock's reading and another clock's reading, but between a clock's reading and the moment in which we find ourselves, the event that is occurring, something that is not ultimately a reading of the clock.[15]

It's important to understand that Bergson was not objecting to Einstein's operational definition of simultaneity: two events are simultaneous in a given frame of reference if they occur at the same time as measured by clocks that have been synchronized using light signals. Rather, he was arguing that the intuitive or experiential concept of simultaneity, which is based on the experience of duration, lay beneath Einstein's operational definition. He was calling attention to the amnesia of experience in the Blind Spot.

Einstein, in his famous 1905 paper, "On the Electrodynamics of Moving Bodies," had claimed to have operationally defined *simultaneous, synchronous,* and *time*, using "certain imaginary physical experiments."[16] At the end of the "Definition of Simultaneity" section, he wrote, "The 'time' of an event is that which is given simultaneously with the event by a stationary clock located at the place of the event, this clock being synchronous, and indeed synchronous for all time determinations, with a specified stationary clock."[17] This definition uses local simultaneity between a local event and a local clock to define the simultaneity of clocks at a distance.[18] But local simultaneity depends on the direct experience of perception. For example, a physical event, such as a ball hitting the ground, is said to be simultaneous with a local clock reading because the two are perceived to happen at the same time. Thus perception, the direct perception of the now as given in the experience of duration, underlies Einstein's definitions. So the definitions cannot claim to be completely operational ones, in the sense of using only objective procedures or tests, and not depending on subjective experience for their meaningfulness.

Bergson seized on these points when he said that the philosophical significance of Einstein's concepts remained to be determined. Einstein's concepts purported to be entirely objective and operationally defined, but our intuitive concept of simultaneity based on the experience of duration lay buried and forgotten in them. Thus, the amnesia of experience in the Blind Spot appears again.

Bergson was not objecting to the need to abstract away from the experience of duration, which always embraces a span of time, in order to conceptualize an exact instant for the purposes of defining simultaneity precisely in physics. He was objecting to surreptitiously substituting the instant for duration and to forgetting that the concept of the instant remains beholden to the direct experience of duration for its meaningfulness.

Einstein appears to have been aware of some of the issues raised by appealing to the perception of local simultaneity. In a footnote in his 1905 paper, he wrote: "We shall not discuss here the inexactitude which lurks in the concept of simultaneity of two events at approximately the same place, which can only be removed by an abstraction."[19] This inexactitude, however, concerned what counted as local versus distant. It didn't obviate the need to rely on the experiential concept of simultaneity as given in the experience of duration in order to abstract a concept of the instant and construct a precise definition of simultaneity for physics.

For Einstein, the final test was whether his theory worked. Allowing for the experience of duration would not have helped him in his theory, so he deemed it irrelevant and ignored it. There is nothing wrong with that. His mistake was to think that his definitions were more fundamental than the experience of time.

Bergson fastened onto these points in his remarks before the society. The inexactitude of the local-versus-distant distinction is the point of both his example of a "superhuman with giant vision [who] would perceive the simultaneity of two 'enormously remote' instantaneous events as we perceive the simultaneity of two 'neighboring' events" and his example of "intelligent microbes," who would find the distance between Einstein's neighboring events (a train pulling into the station while he looks at his watch) enormous:

> They would build microbial clocks which they would synchronize by an exchange of optical signals. And when you would come to tell them that your eye purely and simply observes a simultaneity between event E and clock reading H, which is "neighboring," they would answer you: "Oh no! We do not admit that. We are more Einsteinian than you, Mr. Einstein. There will only be simultaneity between the event E and the reading of your human clock H, if our microbial clocks, placed at E and H, mark the same time; and for an observer external to our system, this simultaneity will be succession, it will have nothing intuitive or absolute.[20]

Bergson's point was twofold. First, what counts as local versus distant is inexact and relative, and depends on the observer. Second, and more important, the ability of the scientific microbes to measure simultaneity by synchronizing clocks using light signals would presuppose their own experience of local simultaneity in duration. The experience of duration—which is to say experience itself—is ineliminable, and cannot be removed by Einstein's operational definitions. On the contrary, those definitions presuppose it.

Bergson was hardly denying that Einstein's definitions were meaningful and scientifically productive. Instead, his argument was that it remained meaningful to speak of simultaneity as given in duration even in the absence of being able to measure duration, and that Einstein had gone too far in claiming to have given an exhaustive, objective definition of time—to have said what time is or what makes time be time. Measurable time presupposes duration, and duration always escapes and exceeds measurement (for the reasons given in chapter 3). This is the philosopher's time that Einstein refused to recognize. For Einstein, the experience of duration was just a contingent psychological fact about how we happen to perceive time given our senses, rather than the ineliminable source for the meaningfulness of the concept of time, including in physics. In Robin Durie's words, "For Einstein, there is no philosopher's time; but for Bergson, the physicist's time is no time at all when it is separated from duration."[21]

Far from wishing to contest relativity theory on scientific grounds, Bergson wished instead to complete it by showing how it required the concept of duration for its proper experiential grounding. What Bergson discerned in the theory of relativity was the possibility, for the first time in modern science, of having "a theory of time which, when interpreted appropriately, discloses the genuine relation between duration and measurable time."[22] Bergson saw that the revolution wrought by Einstein opened up the prospect of moving physics beyond the Blind Spot. A careful philosophical analysis of relativity theory would show that the meaning of measurable time was inextricable from the experience of immeasurable duration, that clock time was inextricable from lived time. Showing this was the task Bergson set for himself in *Duration and Simultaneity*. Unfortunately, this message got lost in the ensuing debate as a result of some mistakes in Bergson's statements about special relativity.

How Many Times

Bergson's main concern in *Duration and Simultaneity* was the apparent conflict between the immediate experience of duration and the plurality of times in special relativity theory. A bunch of issues are tangled up in this "confrontation" between Bergson's and Einstein's theories, so we need to disentangle them to make sense of the debate.

Bergson had argued in *Time and Free Will* that there is one universal time of duration—a single, all-encompassing duration in which all consciousness participates.[23] Nevertheless, the claim that clock time presupposes lived time or that measurable time is an abstraction derived from duration does not obviously entail that lived time or duration is singular, that there are not many durations. Indeed, Bergson himself had emphasized in other writings that there is a multiplicity of durations.[24] Whitehead had also argued that "the measurableness of time is derivative from the properties of durations," but that there are "different families of durations."[25] Whether duration is one or many is a distinct question from whether clock time presupposes lived time. We'll come back to the question about duration being one or many at the end of this section.

What's important now is that Bergson sought to reconcile his belief in a singular and universal duration with the plurality of measured times and time dilation implied by special relativity theory. Time dilation is the difference in elapsed time as measured by two clocks due to their relative velocities. The faster clock is the one at rest and the slower one is the one in motion. But there is no absolute state of rest in relativity theory. Any observer can regard themselves as at rest while regarding other reference frames as being in motion. Time dilation always affects the clock of the "other" observer regarded as being in motion relative to the one taken as being at rest. Bergson reasoned that since there is no absolute frame of reference and the reference frames are inertial (they do not undergo acceleration) in special relativity theory, the observers' situations are symmetrical and interchangeable, and therefore the plurality of times should be regarded as mathematical rather than physically real. And if the many times were regarded as strictly mathematical, then they could be made consistent with there being one real time of duration.[26] This is where he was mistaken.

Bergson focused on the so-called twin paradox. One twin remains on Earth while his brother travels to outer space in a rocket ship at near the speed of light and then returns to Earth at the same speed. According to special relativity theory, when they compare their clocks (which were identically constructed and synchronized at the start of the journey), more time has elapsed for the twin who stayed on Earth, and he appears to be biologically older than his brother.[27] Bergson, however, denied that this would be the case or that this was the right way to interpret special relativity theory. As long as the twins' situations were strictly identical and there was no acceleration (thereby abstracting away from the acceleration required to change direction and return to Earth), he argued, the returning clock would show no slowing on its arrival back on Earth.[28] (In other circumstances in which the twins' situations were different, there would be differences in their times.) In Bergson's view, the clock times were merely mathematically attributed and not physically real.

Here Bergson was incorrect. Special relativity theory does predict that the traveling twin's physical clock has marked less time when the twins compare clocks and that the traveling twin will have aged less, as a consequence of the theory's postulating that space-time has a Minkowski geometry rather than a Galilean one.[29] In addition, time dilation has now been experimentally confirmed as a physical phenomenon.[30]

Bergson argued for two things, one incorrect and the other correct. First, time dilation is not physically real. This was a mistake. Second, no one experiences the time dilation of their own reference frame. The time dilation exists only relative to another reference frame and can be seen only from the outside. This means that time dilation is not a measure of anyone's time from within their own reference frame. A reference frame is not a domain of experience.[31] Instead, it's an abstraction and can be specified only relative to another reference frame. This was not a mistake. Each twin experiences only his own time.

Bergson insisted that the traveling twin does not subjectively experience any time dilation, that he doesn't feel time slowing down for him. To discern that time dilation occurs, one has to stand outside his reference frame and compare clock readings. One might object that Bergson ignored the dependence of experience on brain activity, which also slows down in the traveling twin relative to the stay-at-home one.[32] Hence, the traveler's stream of consciousness elapses more slowly relative to his Earth-bound brother's. Nevertheless, this isn't subjectively noticeable to the traveler. The slower rate of passage exists, or is what it is, only relative to the other reference frame on Earth. So it makes no sense to say that the twin "experiences" a different time. Their experiences of duration remain particular to them. In Bergson's words, a twin who "experiences a different time" is a "phantom," a "mental view" or an "image," appearing to the perspective of the twin on Earth.[33] Here Bergson was correct. Where he went wrong was to infer that the time dilation calculated from outside the reference frame undergoing it is not real and that the two twins cannot really be at two different times as measured physically.

Bergson thought that a measurement of an interval no longer qualifies as a measurement of an interval of *time* if it loses its connection to duration. This is what he thought happens with the different times of special relativity. For Bergson, there was no duration in time dilation. Time dilation shows up only as a difference between clock readings, or a difference between the worldlines (the unique path of an object, approximated as a point, through space-time) computed by the physicist. But no one experiences the different rate of passage as such. It can't be experienced directly, because as soon as you mentally transpose yourself to the reference frame where time dilation is happening, time dilation disappears and reappears in your original reference frame. In Bergson's words, "The times of special relativity are so defined as to be, all but one, times in which we do not exist. We cannot be in them, because we bring with us, wherever we go, a time that chases out the others, just as a pedestrian's lamp rolls back the fog at each step."[34]

Again, Bergson was both right and wrong. He was right that time dilation is computed from outside the reference frame and not subjectively experienced. But he was wrong that "proper time"—the time measured by a clock following along the worldline within a reference frame—cannot be based on the duration proper to that reference frame. On the contrary, proper time can be understood as measurable time based on local passage or duration. Thus, as several philosophers have recently suggested, Bergson's idea of duration is entirely compatible with special relativity theory, if duration is understood as the passage of local or proper time.[35] Moreover, when proper time is understood as a mathematical abstraction based on the experience of duration, now understood as local passage, it becomes possible to reconcile the perspective of special relativity theory with the experience of duration.

Physicist David Mermin makes essentially this point too in a discussion of time and the problem of the now from the perspective of QBism (which we discussed in the previous chapter):

The problem of the Now arises from the long-standing exclusion from classical physics of the experience of the perceiving subject, along with an inappropriate identification of the formalism of physics with the reality of the natural world. That there is such a thing as the present moment is an undeniably real part of the experience of every one of us. The fact that we have a useful formalism that represents our experience and seems not to contain a Now does not mean that Now is an illusion. It means that we must not identify the formalism with the experience that the formalism was constructed to describe.[36]

Let's return, now, to the question about whether duration is many or one. The answer is both.

Duration is many because it is always situated: the passage of time is always given from some experienced perspective in the universe and never from outside it. Each worldline reflects a unique passage and possible experience of duration. Better still, each worldline represents the distillation of a unique durational flow, since a worldline is a mathematical abstraction, a "structural invariant," whereas duration is concrete. Given that there is no nonarbitrary upper bound on the number of worldlines and associated proper times, the universe is teeming with times and potential durational rhythms.[37]

The crucial point that bears emphasizing is that "there can be no such thing as a temporal bird's-eye view of the universe that disregards the discordant rhythms of passage constituted by the disparate paths through space-time."[38] This point refutes blind-spot objectivism—the assumption that there can be such a bird's-eye (or God's-eye) outside view.

But duration is also one. Measurable times and durational rhythms may differ, but temporal passage itself is immeasurable and resists analysis and explanation. Measured time always presupposes the same ineliminable fact of experienced duration or temporal passage. In philosophical jargon, duration is an example of "facticity," something that must be accepted but for which no ground or reason can be given. In Buddhist terms, duration exemplifies "suchness," a concrete character of being for which no conceptual ground can be given. As Whitehead said, nature, as the terminus of sense awareness, is a process, and "there can be no explanation of this characteristic of nature," though we can single it out and describe its relations to other characteristics.[39] Every measurement of a temporal interval is a measure of elapsed time, elapsed time presupposes duration, and you cannot reconstitute duration out of measured time.

In retrospect, we can see that Bergson and Einstein misunderstood each other and expressed themselves poorly in their exchange. This was unfortunate. Combining Bergson's and Einstein's insights can help us to supplant the Blind Spot. As we'll argue now, this move is particularly needed in the case of the blind-spot conception of relativity theory as supporting a "block universe" picture in which time is an illusion.

A Timeless Block Universe?

In the special and general theories of relativity, Einstein showed that time's flow rate may differ from one observer to another, depending on their relative state of motion at constant speeds (special relativity) or due to differences in their local gravitational field (general relativity). Time advances resolutely for every observer, even if more slowly, relatively, for those moving at faster speeds or in strong gravitational fields. The time of relativity remains ordered, as cause precedes effect. Nevertheless, as for Newton, the equations of relativity remain ambiguous with respect to a unique direction of time. They don't distinguish past from future. As we move from Newton to Einstein, time loses its absolute flow rate but remains silent with regard to what differentiates past from future. Mechanics, relativistic or not, has no memory. Duration, however, the time of experience, essentially includes memory, as Bergson said.

Much confusion arises from the fact that Einstein's theories of relativity can be formulated in a four-dimensional space-time continuum, where time seems to acquire equal status to space. But such equivalence is not the case at all. The freedom of motion in space—we can walk anywhere we want, north or south, east or west, up the mountain or down the valley—is not reflected in the time dimension. There, we are still constrained by the causal chain, where events succeed one another in an ordered fashion. You can stop walking and stay at the same position in space; you cannot stop living in time.

Mathematically, and for a flat geometry, the difference between space and time is encapsulated in the four-dimensional Minkowski space-time where the square of the invariant distance ds between two events in space-time is $(ds)^2 = -c^2(dt)^2 + (dX)^2$. In this equation c is the speed of light, t is time, and X is position in three-dimensional space. (The d's indicate that these are tiny segments.) The time variable t appears multiplied by the speed of light, since "c multiplied by t" has dimensions of space. The sign difference between spatialized time and the three space dimensions is crucial.[40]

In Minkowski space, the distance ds between two events (taking place at two points in space and two moments in time) is the same for all inertial frames of reference. (This is what one means by "invariant distance in space-time.") Causality is built into this formulation implicitly, safeguarded by the speed of light, because the square of the distance between two events cannot be negative. Mathematically, $(ds)^2 \geq 0$. The smallest possible distance, $(ds)^2 = 0$ (the null, or light cone separation), is realized only for $|dX/dt| = c$, that is, when an "object" is moving at the speed of light. The theory also determines that only massless "objects" can do so. Every other speed, being smaller than the speed of light, will render the square of the invariant distance $(ds)^2$ positive. This is how time ordering is built-in.[41]

What seems arbitrary in this framework is the definition of the present, or the "now." One can choose any time to be the now, the moment when the clock starts ticking. Both the special and general theories of relativity added a locality to this arbitrariness that was foreign to the Newtonian worldview: the special theory, with the arbitrariness given by the relative motions of observers moving with constant speeds; the general theory, with local distortions on time flow due to variations of the gravitational field, that is, different time flows in regions with different concentrations of mass.

This apparent arbitrariness of time's flow and the meaning of now have inspired some scientists and philosophers to propose that there is no ever-changing now. Instead, all change is illusory.[42] In this way, they use theoretical tools from Einstein's relativity theory to echo pre-Socratic philosophers like Parmenides and Zeno. Going by the name of eternalism, the core notion is that just as the diagrams that display the whole of space-time seem to reflect a timeless reality of being, it is our narrow three-dimensional view of reality that brings forth notions of past and future. In the full glory of four dimensions, there is no time flow. This view is often called the block universe theory: all of space-time is an unchanging four-dimensional block.[43] Accordingly, all cosmic history and the entirety of the future constitute a single block in four-dimensional space-time, and our experience of time's flow is illusory. In the words of mathematical physicist and philosopher Hermann Weyl, "The objective world simply *is*, it does not *happen*."[44]

The block universe theory epitomizes the Blind Spot. How is it possible to formulate coherently a conception of the contemporaneous reality of all the events contained in the block universe without adopting an impossible God's-eye perspective external to the universe and the passage of nature?[45] Besides being objectivist, the block universe theory bifurcates nature into objectively real space-time and illusory psychological time, surreptitiously substitutes space for time (spatialized time for duration), and treats an abstract mathematical representation as if it were concrete being (the fallacy of misplaced concreteness). The theory robs nature of passage and time of duration. It reifies space-time by turning the process of becoming into a thing. In Bergson's words: "By adding a dimension [time] to the space in which we happen to exist, we can undoubtedly picture a *process* or a *becoming*, noted in the old space, as a *thing* in this new space. But as we have substituted the *completely made* for what we perceive *being made*, we have . . . eliminated the becoming inherent in time."[46] The block universe theory confuses a mathematical picture with what is being pictured; it confuses the map with the territory.[47]

Time's flow is palpable, even if relativity theory shows us that the rate of our flow of time is not universal but rather local to us as observers. Thus, if our goal is to offer a map of reality, we have two options: offer a map that invokes an abstraction to discard the flow of time, or one where the flow of time is an inherent part of our experience and of an unbifurcated nature. What would be the purpose of a map that discards the flow of time? Where does it lead us? Does it help us understand time any better or lead to intractable conundrums?

One of the lessons from our discussion of Bergson and Einstein is that there cannot be a temporal bird's-eye view of the universe, one that flies outside and above the disparate paths through space-time and the different rhythms of duration. The block universe theory renounces this insight, pushes physics back into a blind-spot worldview, and remains stuck with the intractable conundrum of being unable to account for the temporality of time—time's passage, its flow, and its irreversible directionality. For these reasons, the block universe theory is essentially regressive. It reinstates the Blind Spot instead of helping us get beyond it.

Einstein is often said to have believed that time is an illusion, based on a cryptic remark he made in a letter on the occasion of the death of his lifelong friend, the Swiss-Italian engineer Michele Angelo Besso. In a letter addressed to Besso's son and sister, Einstein wrote, "Now he has departed from this strange world a little ahead of me. That means nothing. For us, believing physicists, the distinction between past, present, and future is only a stubborn illusion."[48]

But these remarks are ambiguous. On the one hand, Einstein was not addressing physicists and philosophers.[49] The letter expressed his grief and wish to console Besso's family, and he appeared to be thinking about his own approaching death (he died one month and three days later). On the other hand, he offered solace by drawing on his standing as a scientist to opine about metaphysical and spiritual matters. As a "believing physicist," he knows the true, hidden nature of time, despite the contrary testimony of his experience. We must allow Einstein his consolation, but we should not ignore the existential predicament he faces from treating time as an illusion.

Duration, passage, the flow of time: these are essential to how we experience the world. Science would be impossible without them. Any science that tries to attribute the flow of time to a subjective illusion induces an amnesia of experience and cuts away the ground on which it stands. To imagine that we can abstract away from the flow of time to attain an objective view of reality is to replace the passage of nature with a lifeless symbol. There is a crucial difference between how we can be misled by sensorial experience—which Einstein's theories attempt to correct by focusing on observational invariants—and discarding experience altogether as being irrelevant to our description of reality.

Thus, the task at hand is to reimagine a map of reality that includes, implicitly, a time direction that embraces all length scales, from the atomic and molecular to the human and the cosmic. Indeed, as we suggest in the next section, our own experience of time carries within itself the whole history of the universe.

The Emergence of Cosmic Time

In chapter 3, we saw that thermodynamics attributes the arrow of time to the loss of information that happens as a large number of molecules interact with one another. Tracking down the details of every collision requires a precision of description in phase space that is unachievable, even in principle. This is critical. No measurement can have arbitrary precision. This fundamental limitation compromises the task of mapping reality. Small errors in measurement propagate, information is lost, and reversibility is essentially impossible. To put it bluntly, we can't know everything, even about fairly simple systems. Forgetting these limitations represents another aspect of the amnesia of experience in the Blind Spot, this time of scientific experience and its inherent limitations.

There is a second, equally fundamental, complication of our efforts to map reality, one that doesn't depend on our measuring limitations. Systems with a large number of particles are chaotic: tiny perturbations, even if very far away, affect the system in such a way as to make it impossible to predict its behavior deterministically (the famous butterfly effect in weather prediction). A flea jumping far away in the universe would lead a gas of atoms into chaos after about thirty-three collisions, in under one-millionth of a second. The tiny effect would take a long while to reach the gas, but once it does, the effect is lightning fast.[50] We thus have a conflation of two fundamental limitations: chaos—perturbations that lead to changes independent of us—and the physical limits of precision measurements.

Still, someone might argue, Can't we at least imagine a perfect being or miraculous device that would record every molecule's speed, location, and angle of collision with infinite precision? Such a being would be a synthesis of Laplace's demon, who knows the precise location and momentum of every particle in the universe, and thereby their past and future values for any given time, and Maxwell's demon, who can follow every particle in its course and intervene so as to decrease entropy. Couldn't such a being reverse motions perfectly, and thus decrease entropy and invert the arrow of time? To this being, the past and the future would be equally accessible and controllable.

But these feats remain impossible even for such a being unless three fundamental laws of physics are violated.

First, the being would need to be omnipresent and omniscient: it would need to know the details of all individual measurements of position and momentum across space instantaneously. But how could this be, given that every measurement is an interaction in a region of space and at a moment in time? In other words, the very notion of causality and the finiteness of the speed of light preclude such measurements from taking place in any even remotely realistic sense. Naturalism clashes with supernaturalism.

Second, this being would have to be able to measure the positions and velocities of every molecule and potentially interfere with their trajectories without generating any entropy. But how could it function and do work on the system (manipulating molecules) without outputting entropy? That would violate the second law of thermodynamics.

Finally, there is quantum uncertainty. Recall that Boltzmann's and Gibb's treatment of statistical mechanics requires the definition of phase space, composed of small cells denoting a region of space (like a tiny cube) and of speeds (more precisely, momenta). When we introduced this concept of phase space in chapters 2 and 3, we addressed the coarse graining or blurriness intrinsic to defining the size of a cell in phase space. How small can a cell get? Classically, it depends on the precision of our measuring devices. Arguments like the one above, invoking a magical being capable of measuring physical quantities to arbitrary precision so as to squeeze the size of a cell in phase space to a point, are a fabulation, a theoretical exercise that tries (and fails) to evade the Blind Spot. Eventually, as the size of the cell shrinks, the limits of quantum uncertainty prohibiting measurements of position/momentum with arbitrary precision will start to matter. Quantum blurriness is the ultimate coarse-graining limit.[51]

Coarse graining thus plays a dual role in our understanding of physical reality. There is a blurriness that is intrinsic to the very fabric of the world, a fundamental feature of physical reality at the quantum level; and there is a blurriness that stems from how we interact with the physical world, through the limitations of our measurements and data gathering and storing. If we accept the relation between the arrow of time and loss of information, this means that entropy in closed systems grows and time has a unique direction from past to future.

Can we elevate this notion from molecular or atomic collisions to macroscopic scales? In other words, is it possible to amplify the irreversibility of time from molecular to macroscopic scales? After all, time flows everywhere toward the future, even if at different relative rates depending on the local gravitational field or the relative motion of observers of objects. Time is intrinsic to all living creatures, from unicellular bacteria to thousand-year-old bristlecone pines; stars are born, evolve, and die; galaxies move apart from one another as the universe expands. Time seems to run forward across all spatial scales, from the very small to the very large, from the quantum to the cosmic.

As far as we can tell, there are no places in the universe where time flows backward. This has been true long before humans were making measurements and thinking about time. Due to our coupling with the world through experience, however, what we can tell of the prehuman past depends on the narrative we construct in the present, anchored to what we can measure of nature. Our image of the past resides in the present, and the story we can tell about the past is necessarily incomplete. It remains bound by the limitations we have mentioned. Forgetting these limitations while extrapolating from modern physics to the distant past amplifies the blind-spot errors and misconceptions about time and the limitations of science, as we'll see further below. If understanding the direction of time is challenging, understanding the origin of time escalates the difficulties to a whole other level.[52]

Take as a starting point the following question: What links the arrow of time across all known physical scales, from the subatomic to the cosmological? The link is the very history of the universe as we now tell it, from an initial primeval fireball just after the big bang 13.8 billion years ago, to the expanding space that forged atoms, stars, galaxies, and life on Earth (and possibly elsewhere). The discovery of a cosmic history, aligned with the dating of our own planet at about one-third of the cosmic age, represents a profound shift in our perspective on time. For thousands of years, the regularity of local celestial movements allowed us to order the flow of time in periods of hours, days, and years, timescales that we can relate to, ones not so different from our life span. But modern dating techniques and developments in astronomical observation and spectroscopy in all windows of the electromagnetic spectrum have now extended the cosmic clock to billions of years. The old cosmos of being, a static receptacle for all that exists, turned into a universe of becoming, punctuated by dramatic shifts in the nature of material structures, ranging from subatomic particles to giant galactic clusters and beyond. Those shifts in the properties of matter determine how the universe expands, and thus how cosmic time flows. The kinds of stuff that fill the universe determine, through their gravitational interplay, how fast the universe expands.[53] The greatest discovery of modern cosmology—the expansion of the universe—redefined our own experience of time, which now can be seen as deeply enmeshed with the history of the cosmos itself.

Such statements about the expansion rate of the universe are possible because we interact with the change we measure. Indeed, the very idea of cosmic time as a measure for the evolution of an expanding universe—time as measured by a clock at rest relative to the expanding universe and set to zero at the time of the universe's initial state—is an abstraction. It cannot be made intelligible without ultimately referring back to our experience of time as duration, as Bergson and Whitehead have shown us. Like the fabled ouroboros, the serpent that swallows its own tail, closing itself in a circle, our experience of time depends on the flow of cosmic time that we measure through our experience of time. We cannot extricate ourselves from cosmic history because it is our history too. We are in the universe and the universe is in us. We are caught up in a strange loop.

Cosmological Conundrums

When Edwin Hubble discovered in 1929 that distant galaxies were receding from one another with speeds proportional to their distances, the universe itself gained both a plasticity and a history. Twelve years earlier, Einstein had applied his general theory of relativity to the universe as a whole.[54] But his imaginary cosmos was quite different from what Hubble saw with his telescope. Assuming a spherical three-dimensional universe with no boundary—as we can visualize in the two-dimensional surface of a ball— and a static cosmos, Einstein solved the equations of his general theory of relativity to find the geometry of a self-gravitating spherical universe, launching modern-day relativistic cosmology. To his disappointment, however, the solution he found was unstable against collapse. As a way out, Einstein added an extra term to balance his equations, which became known as the cosmological constant. Five years later, in 1922, the Russian meteorologist-turned-cosmologist Aleksander Friedmann showed that one could remove the assumption of a static cosmos to obtain time-varying solutions that described either a forever-expanding universe or one that would reach a maximum size and then collapse upon itself, cycling from a big bang to a "big crunch."

Skipping to modern cosmology, we now know that the universe has been expanding since its infancy, 13.8 billion years ago. We don't have the details of what truly happened near the beginning, and speculations abound. For example, in classical relativity, there is an unavoidable singularity as one reaches back to the beginning, $t = 0$, given that physical quantities such as energy density and temperature diverge to infinity as the distance between objects goes to zero. To avoid this, quantum effects are called forth to change the narrative as we approach the singularity.

If—potentially—the very structure of space-time would fluctuate, the singularity would literally be blurred away.

It is essential to keep in mind, however, that we don't know, or have any indication from observations, that extrapolating quantum fluctuations to space-time itself is a valid step. Such fluctuations are indeed essential to our current understanding of quantum fields as the fundamental players of material reality (the building blocks of the universe, as some scientists like to call them). Those players are the twelve fundamental particles of the Standard Model of particle physics (six leptons and six quarks), the Higgs boson, and the particles that carry the three known forces of nature (electromagnetism, and the strong and weak nuclear forces). Gravity, negligible at the subatomic distances described by the Standard Model, is added on as the fourth force, with the implicit assumption that, as with the other three forces, it is also quantizable and that its carrier, the graviton, is a massless particle. This is plausible but certainly not a known fact. It's plausible because gravity, like the other three known forces, is described by a field theory. If gravity is a field, why not quantize it?

But this assumption, though plausible, is troublesome. Quantizing gravity is not straightforward, because gravity is inherently different from the other radiation and matter fields. These other fields are defined as having values in space-time, that is, at each point in space and time. When such fields are quantized, space and time remain point-like. But what would a quantum theory of the electromagnetic or the weak and strong nuclear fields look like on a quantized space-time? No one knows, and opinions diverge.

Our current quantum field theories are effective coarse-grained descriptions of how particles interact in space-time, spectacularly useful to describe the results of the experiments we now can perform. The trouble is that such experimental evidence holds at energies fifteen orders of magnitude smaller than those near the big bang. Clearly, any extrapolation of current physics by this amount should be taken with a huge grain of salt. But it often isn't. Practitioners take what we know at our current accelerator energies and tacitly assume that reality will be of a similar nature in a completely different environment, one for which we have no data whatsoever. Through surreptitious substitution, they forget that fields are just models of the world, not the world itself. Field theories are maps, not the territory. Maps can be effective only when they are designed based on known parts of the territory.

Unknown parts are not mappable, hence the "Here Be Monsters" regions of old medieval maps.

Forgetting these cautions leads to blind-spot thinking, particularly to an unwarranted objectivism. Since field theories are effective at the energies we can test, it seems natural to extrapolate them to much higher energies, where we have no data. What else could one do to advance scientific knowledge? But it's dangerous to forget how these extrapolations amplify the limitations that are already problematic with current effective theories. Given that what we can say of physical reality is ultimately and irreducibly based on experience, every extrapolation into a no-data territory needs to be critically scrutinized. If there are no empirical tests to validate the hypotheses, all we can hope for is compatibility with the parts of the universe we can test at much lower energies. Thus, it's hubris to consider such theories final in any genuine sense, even though this is often done. To declare that we have, or are close to having, a final understanding of the subatomic world and of the quantum nature of space-time is bad science and bad philosophy. This way of thinking illustrates how lost the search for order in nature can become once blind-spot objectivism holds sway.

In practice, applying the same recipe used to quantize the three force fields to the gravitational field leads to severe technical dead-ends. Attempts to move beyond them—in particular, current candidates for a quantum theory of gravity such as loop quantum gravity and superstrings—have not proven compelling, despite their beautiful formal structure and over five decades of focused efforts by thousands of talented physicists. In particular, the lack of detection of any hint of supersymmetry, an essential ingredient of superstring theory, has left its candidacy as a fundamental unified theory of reality on shaky grounds.[55]

Of course, failure should not be a deterrent for further efforts. We find only if we look. But it's precisely in these situations where it's tempting to extrapolate that we must proceed in full cognizance of the Blind Spot. There are fundamental limitations to how far our theories can go as we attempt to describe reality. As we've argued, any theory that tries to extricate itself from our experience of the world, extrapolating far beyond perceived and measured reality to propose a God's-eye view from nowhere and nowhen structure of space and time, radically alien to how we are in the world, will end up being a ghost chase. There is only so far one can go without data as a guide. Such extrapolation, added to an expectation of explanatory finality, sets up in advance an objectivist conception of reality severed from the act of knowing. This is a mistake. Science is based on what we perceive of the world, so attention to the act of knowing and its limitations must temper any conception of reality. Decades ago, Isaiah Berlin called the insistence on the truth of final explanations the "Ionian Fallacy," referring to the ancient Ionian tradition of natural philosophy that began with Thales who said that all is water (see chapter 2).[56] Perhaps gravity is simply different from the other forces and cannot be consistently formulated at very short distances. Many in the physics community are beginning to think this may be the case.[57]

Getting Things Going

So what can we say with confidence about the history of the universe? Quite a lot actually. Fortunately, to build a basic narrative of time at the cosmic scale, we don't need all the details of what happens at or very near the big bang. What we do need to know is that the universe has a history and that this history, like any other, unfolds in time, directly or indirectly affecting all physical scales. Every material structure that exists, from atomic nuclei to clusters of galaxies, from stars to planets to living creatures, emerges and is sustained by a thermodynamic exchange of available free energy into entropy. For ordered structures to exist, free energy must be available so that work can be done, inevitably leading to entropy growth, as the second law of thermodynamics dictates. What this means is that the entropy of the universe in the far past had to be smaller than the entropy now. Otherwise, no structures such as stars could have formed, and hence no warm, water-rich planets or living creatures. A high-entropy universe would be filled with a formless gas of radiation and little else. We could even say that life is what eventually happens to an old, hydrogen-rich universe that begins in a low-entropy state. At least that's our story.

Current experiments at the European Center for Particle Physics (CERN) are probing physics at one-trillionth of a second after the big bang. At that time in cosmic history, the universe was filled mostly with what is well approximated by a radiation fluid, composed of massless particles (like photons and, potentially, gravitons) and effectively massless elementary particles (like neutrinos and many others).[58] This primordial fluid was gravitationally "unclumped": there were no stars, no galaxies, no self-gravitating bodies.[59]

How far back we go toward this low entropy past depends on both our evolving understanding of the physical processes that took place very early on during cosmic infancy and our tolerance for speculative theoretical physics. For practical purposes, we don't need to go too far back. Instead, we can explore two crucial events in the cosmic history.

The first great cosmic event occurred around 50,000 years after the big bang, when the universe transitioned from being radiation dominated to being matter dominated. In other words, the massive particles that existed took over massless radiation (such as photons and neutrinos) to determine how fast the universe would expand.[60] If, at this time, dark matter—unknown particles that contribute to the composition of the universe—moved with speeds well below the speed of light (so-called cold dark matter), gravitational attraction would have caused mostly dark matter-rich clouds to start clumping. These clumps were the seeds of what would become stars, galaxies, and clusters of galaxies hundreds of millions of years later. Think of many kids sliding into a dugout pit: the dark matter clumps carve the gravitational pits, and ordinary matter slides into them. Gravitational clustering is a slow process that uses potential energy to squeeze matter into smaller volumes. The tighter the squeeze, the faster that particles and other objects move, causing an increase in entropy.

The second great cosmic event happened at about 400,000 years after the big bang, when photons decoupled from electrons and protons to become what today is the microwave background radiation. The protons and electrons joined to form the first hydrogen atoms—matter as we know it. These atoms slowly drifted toward the gravitational wells dug by the dark matter clumps and started to get squeezed. Since the photons had now mostly decoupled from matter, there was almost no radiation pressure to counteract the gravitational squeezing.[61] At about 200 million years after the big bang, the first stars were born.

The rest is history, albeit complicated and dramatic. All the while, free energy is bootstrapped into progressively more complex material structures as entropy increases in fits and starts. Planetary systems emerge and chemistry explodes into countless possibilities until, in at least one world and potentially more, molecular assemblies capable of self-production and agency metabolize energy and reproduce themselves while increasing the entropy around them (see chapter 6). Some of these, we may suppose, evolved to be able to think and ask questions about their origins, their curiosity driven by an acute awareness of time's passage given in the experience of duration.

Confused Beginnings

For the narrative just described to work, however, a crucial step is required. We must assume that the universe started in a low-entropy state in the far past, as encapsulated in a uniform radiation fluid. Radiation has lower entropy than clumped matter, especially black holes. But is this assumption reasonable? Clearly, if we consider many possible beginnings for the universe, examples with higher entropy are much more common. A low-entropy initial condition is a rare event, but one that we need so that our current narrative makes sense.

This need for a rare low-entropy initial state has been called the "past hypothesis."[62] There have been many attempts to address the issue in theoretical models. One route has been connecting the beginning of time to a phase before the big bang where time ran backward, a sort of mirror universe.[63] Another possibility is considering that the laws of physics may have varied very early on, before the universe settled into its current course.[64] Researchers have also considered that the universe goes through a succession of expanding and contracting phases, reaching a sort of bounce state that never really squeezes space down to a problematically small size.[65]

To date, none of these models has been broadly accepted by the scientific community as a solution to the low-entropy problem. From the perspective of our critique of the Blind Spot, we can see why. The three attempts listed above extrapolate current physics to the beginning of time itself. Other models do the same, even though we can't trust any current extrapolation to very early times, given that we don't have sufficient empirical information to ground our speculations. Of course, we understand that the point of speculation is to push the limits of current knowledge to situations where it may or may not be valid, so as to gain some level of understanding of viable hypotheses. Speculate we must.

Nevertheless, proposing models of the cosmic beginning implicitly assumes that the mathematization of time as a continuous (or discrete) variable will be of use as we go down to $t = 0$ or very close to it. The modelers assume that the same physical laws that apply at the limit of our current experiments at high energies will apply as we move back closer to the beginning of time, where energies are thousands of trillions of times higher. Ironically, current scenarios to address the mystery of time's arrow sometimes use our ignorance of the nature of time near the cosmic singularity to propose solutions to the problem of time: the Blind Spot becomes a guiding principle to trace new ideas that lead, as one might expect, to dead ends.

The multiverse (a hypothetical group of multiple or parallel universes) is an example. There are essentially two different ways to think about the multiverse in cosmology: one from the so-called inflationary models and the other from superstring theory.

In the case of inflation, it is assumed that the primordial universe was filled with a scalar field that determined the cosmic expansion rate. "Inflation" refers to a short-lived period of exponentially fast expansion where this scalar field, often called the inflation, is displaced from its energy minimum. This excess energy translates into an effective negative pressure that drives the exponential expansion. The details are unimportant here. The multiverse emerges from this picture when one adds quantum effects to the dynamics of the inflation. Picture the field rolling down its energy curve like a child going down a slide. But since this is a field in the early universe, quantum fluctuations are important and may kick the field both farther down or farther up the energy curve. The more energy the field has, the faster the universe expands. As a result, in different regions of the universe, this field will be doing different things. In some regions, it will be kicked down, while in others it will be kicked up. The net result is a proliferation of cosmic regions, each expanding at a different rate, each generating a different kind of universe. The sum total of these regions, an ever increasing number of them, is the multiverse in inflationary cosmology. We are supposed to inhabit one of them, forever separated from our neighboring cosmoids.

In superstring theory, the multiverse emerges from considering that these theories are formulated in spaces with six extra spatial dimensions. These dimensions are curled up into a tiny space, with complicated geometry. The effective theory that we "see" in our ordinary three spatial dimensions is directly determined by the details of the geometry of these extra dimensions. A different geometry "there" means a different kind of physics here. The string multiverse emerges as a form of landscape of possible geometries. Each universe originates from an extra-dimensional space with its own particular geometry. It turns out that this geometry also determines the values of the physical parameters in our universe, such as the mass of the electron or the quarks, their spins, and so on. So each universe in the landscape will have a different kind of physics. Ours happens to be the one that has the values that are ripe for a long-lived universe filled with stars, and at least here in this planet, life. Proposers of this kind of multiverse claim that it answers a fundamental question about our universe: Why is it "tuned" for life? It isn't, they claim; it's just that we "won" the cosmic lottery.

Let's consider these two multiverses under the lens of our critique of the Blind Spot. The inflationary multiverse relies on a massive series of extrapolations: that a scalar field existed at very early times; that it was modeled with an energy profile of a specific kind; that the small entropy condition was satisfied; and that quantum effects did indeed do what the model claims they are capable of doing. Although compelling, none of these assumptions rests on any firm empirical ground. They are extrapolations from the known conditions in current experiments, far removed from the furnace of a universe at 10^{36} seconds after the bang. Variations on this simple scenario call for similar extrapolations. It is not wrong, of course, to extrapolate, as we have stressed before. But it is wrong to build grand explanatory narratives with a surprising level of confidence based on such extrapolations. That's objectivism and surreptitious substitution in action. That the model achieves a certain level of consistency with current observations (the so-called Lambda-CDM model) is a far cry from its validation. Consistency is the least that you can expect from an extrapolation; it's no proof that the extrapolation is valid as a model of physical reality.

The string multiverse is even more drastic. Here we not only have the extrapolation of fields with behaviors that are familiar to us from conditions that are prevalent much later on, but also the following added assumptions: (1) space has six extra dimensions; (2) supersymmetry, a hypothetical new symmetry of nature that doubles the number of particles, is valid; (3) the ontological substrate of physical reality is neither particles nor fields but vibrating strings; (4) the many topologies of the extra six-dimensional space indeed correlate with different kinds of universe in the landscape, ours being one of them; and (5) there are only four fundamental forces of nature, and they can be unified into a single force under this scenario.

The string multiverse combines several misguided blind-spot ideas, most clearly the fallacy of misplaced concreteness—the fallacy of mistaking a highly abstract, mathematical, speculative, theoretical edifice for nature itself. The string multiverse is a direct heir of Platonism in modern physics. It epitomizes the belief—for it is a belief—that mathematics is the blueprint for the secrets of a "monotheogeometric" description of physical reality: symmetry is understood as beauty, and beauty in this mathematical sense is the way to the truth. String theory also epitomizes the spiral of abstraction that characterized classical mechanics as it moved from Newton to Hamilton and Poincaré, but with the fundamental and critical difference that its postulated reality is far removed from empirical validation. String theory is the apex of the Blind Spot in modern physics, an illustration of the confusion between mathematical abstraction and nature that Whitehead admonished against.

In the future, we're likely to gain more empirical information about the universe's distant past, the fossils of cosmology. These data will certainly help to ground our models attempting to build a narrative of physical conditions near the beginning. Nevertheless, we cannot realistically expect to have a bridge linking our experience of time, the essential ingredient in all measurements of time, to the initial conditions that set the universe on its course. Such a bridge would need to take us outside the realm of time and space, outside the universe, beyond science as we know it, giving us a God-like perspective that we have seen to be a fiction.

We conclude this chapter with a few words on the question of time's beginning. Cultures from across the globe have been fascinated with the origin of all things and have created mythic narratives that attempt to make sense of what are perhaps the hardest questions we can ask: How did the universe begin? What is the origin of time? Why is there something rather than nothing? Creation myths are a bridge from the unknowable to the knowable. Many of them link the idea of the cause that started all causes to supernatural intervention. They posit a first cause in the form of a God or deity. Other traditions see the universe as beginningless. We see a similar divergence in later systematic philosophies. Thus, whereas Aristotle introduced his eternal unmoved mover as the first cause of the single causal system of the universe, medieval Indian Buddhist philosophers such as Dharmakīrti and Ratnakīrti argued that the very idea of something's being eternal while also being able to act as a cause, or more generally the idea of a first cause that was not itself an effect of something else, is contradictory.[66]

Science, of course, cannot use supernatural intervention or a priori arguments about the necessity of a first cause, so it hits a hard wall with the problem of the first cause. Still, some scientists have claimed to have solved the problem of the first cause by adding quantum effects to classical space-time in a "quantum cosmology." The models are elegant and captivating, but they have no explanatory power. They are all a consequence of the Blind Spot, of elevating many working physical assumptions to the level of axioms, and building fabulist maps that are more like useful interpretive devices than genuine maps of reality. Many concepts are needed to build such quantum cosmology models—fields, energy, space-time, conservation laws, Hamiltonians, mini-superspace, Wheeler-DeWitt equations—and none them is justified a priori. Models are maps, and maps work well only when the mapmakers know the terrain they are mapping. Quantum cosmology models live within the "Here Be Monsters" regions of the map. We can either accept that they are fabulations or, alternatively, accept that certain questions transgress the bounds of science. Who decreed that science must be able to answer all questions about reality? If what we can say of the world depends on our experience of the world, to describe what lies beyond any possibility of experiential confirmation belongs to the realm of gods, not people.

III   Life and Mind

6 Life

Life in the Blind Spot

Life is a surprise, at least from the perspective of physics and chemistry. If an advanced AI system were given all the laws of mechanics, all the equations for electromagnetism, and all tables for all possible chemical reaction chains, it is still not clear that it would ever stumble on the kinds of systems we call "living." Even worse, if it were presented with such systems, it's not even clear that it would recognize them as living ones. Knowing only the laws of physics and chemistry, it might simply be unable to recognize that the living systems—the realm of the biotic—are different from the nonliving ones—the realm of the abiotic.

Our imaginary AI system is not alive. It's a purely computational intelligence, with no living body and no metabolism, and hence no need to constantly renew and maintain itself. Its knowledge is limited by its having a certain kind of world model—what AI researchers call a "cognitive ontology," a specification of what kinds of things exist in the world and the relations among them. Physics and chemistry exhaust its ontology. How would such a lifeless, disembodied AI system be able to recognize life? How would it be able to tell which phenomena are living beings? How would it be able to tell that the heart's function is to pump blood and not to make sounds, since both are equally physical effects? How would it be able to tell the difference between a healthy process and a pathological one, since the laws of physics and chemistry don't differentiate them? As French philosopher and physician Georges Canguilhem writes, "Whereas monsters are still living beings, there is no distinction between normal and pathological in physics and mechanics. The distinction between the normal and the pathological holds for living beings alone."[1] In short, how would our AI system, being confined to a physicochemical representation of the world and lacking the concept of a living being, to say nothing of the experience of being alive, be able to bring organisms into focus, let alone distinguish within them between function and side effect, or health and disease?

But *we* know there's life. We experience our living bodies from within, and we recognize other bodies as alive. This experience happens long before we learn biology and is a condition of possibility for biology, even though biology enriches and revises it. Certainly we've greatly expanded our perception and comprehension of the living world, thanks to biology and the tools and methods of the scientific workshop. Still, only life can know life, as philosopher Hans Jonas writes.[2] The biologist can validate our experience of life after the fact, and can deepen and emend it, but cannot forswear it. This is true not just in practice but also in principle: as Jonas argues, a disembodied, mathematical physicist-chemist god would be incapable of delimiting living beings in the physicochemical flux.[3] It would perceive only the flux of atoms and molecules, but not the individuality of the organism and how the organization of the organism channels the molecular flow. Without the experience of embodiment and the sui generis concept of life, such a god would have no way to recognize the standing form or pattern that demarcates an organism as a persisting individual in the incessant turnover of molecules, atoms, and elementary particles.

This strange loop—that it takes life to recognize life—is what the Blind Spot hides when it comes to biology. Just as the science of thermodynamics presupposes our bodily experience of heat, so the science of biology presupposes our experience of life. Our perceptual experience, not physicochemical or functional analysis, is what enables us in the first instance to discern physical movements as behaviors—as swimming in fish or bacteria, flight in birds or insects, or growth in plants—even if our perception sometimes is mistaken in attributing life to nonliving artifacts. Biology presupposes life's experience of itself. The amnesia of this experience defines the Blind Spot conception of biology.

In this chapter, we turn from the physics of matter and time to the biology of life. Science has made extraordinary progress in understanding the molecular basis of life through the discoveries of DNA, RNA, and protein synthesis. Much of that progress has been based on applying the method of microreduction (analyzing a whole in terms of the properties of its parts). As a result, the Blind Spot perspective on life leans heavily on the metaphysics of reductionism (the idea that the properties of complex systems are determined exclusively by the properties of their parts). For reductionist metaphysics, life is nothing more than molecular machinery. This way of thinking is a case of surreptitious substitution, of replacing a sometimes useful method with a metaphysics.

The life-as-machine metaphysics, however, is deeply flawed. The machine metaphor works only for parts abstracted from their context but not for self-producing and self-sustaining wholes. Organisms, as organized whole systems, produce themselves, repair themselves, and generally maintain themselves. "Biological autonomy" is the name for this kind of systemic organization.[4] Self-individuation, agency, and dependent autonomy—"needful freedom" in Jonas's words—make life unlike anything else in nature or anything we manufacture. The life-as-machine metaphysics does not capture this unique mode of being and way of being organized and thereby fails at its explanatory task.

Besides reductionism, other hallmarks of the Blind Spot, particularly physicalism and epiphenomenalism, appear in mechanistic treatments of life. At their root stands the bifurcation of nature, which demands that life's most salient features, such as autonomy and agency, be split off as epiphenomena of more fundamental molecular processes. Narrowly defined parts, such as genes abstracted from the developmental context of the organism and its environment, are elevated as the essence of life. Such parts are surreptitiously substituted for the whole, the self-producing and world-constructing organism, leaving it hidden and unexplained.

Our focus will be on how advances in biology demand that we supplant the Blind Spot conception of life as a molecular machine with a perspective on life that emphasizes biological autonomy and open-ended historical becoming.

Fires, Hurricanes, and Stars

Despite spectacular advances in biochemistry and genetics, there is still no consensus on how to define life or on exactly how to characterize the difference between living beings and nonliving physical systems. Consider three different physical systems: fires, hurricanes, and stars. All three have the general thermodynamic properties of what are called "dissipative structures"—systems that operate far from thermodynamic equilibrium by exchanging matter and energy with their surroundings. They generate large-scale patterns (such as the cyclone of a hurricane) as the energy flowing through them displaces them from equilibrium. Although dissipative structures abound in living systems, being a dissipative structure isn't sufficient for being an organism: fires, hurricanes, and stars are not organisms. Highlighting the differences between these systems and organisms can help us think about what life is and what it does.

Consider fires first. Fires spread and feed on their environment to sustain themselves. They consume oxygen to keep on burning, and thus are open thermodynamic systems, like living beings. Given the right conditions, fires multiply, often with devastating consequences. But we know that fires are different from organisms. We wouldn't call oxygen combustion a metabolic process. Unlike a cell, a fire does not have an internally produced physical boundary. In geometrical terms, a fire has a changing physical shape but no topological unity thanks to a membrane. We also wouldn't consider the spreading of a fire to be a form of reproduction. One fire does not duplicate its structures and divide into two, as happens in cell division. Nor does a fire replicate: it doesn't serve as a template for creating another fire, as happens in DNA replication. For these reasons, fires don't have an evolutionary history. They don't form lineages with conservation of and variation in structural characteristics from one generation to another. Finally, fires aren't goal directed. If a fire is burning down a ravine toward a creek, it will keep on burning until it stops by the water and eventually dies out. It won't forage for more fuel in order to keep burning. A cell, however, adaptively regulates the flow of matter and energy so that it remains in a far-from-equilibrium state while avoiding conditions deleterious to it and altering the environment in a way conducive to its own continuation.

What about hurricanes? Like fires, they are persistent, far-from-equilibrium complex systems that need the right environmental support to exist and maintain themselves. They travel and are tightly coupled to local humidity, pressure, and temperature conditions. As long as favorable atmospheric conditions hold, they maintain their basic shape. Jupiter's Great Red Spot is a giant anticyclonic storm that has endured for at least four hundred years. But as with fires, we wouldn't equate these properties of hurricanes with being alive in the way a cell is alive.

Stars are similar. They are self-sustaining in the sense that they convert gravitational potential energy (they are slowly imploding) into the enormously high pressures and temperatures at their cores that promote the nuclear fusion reactions needed to sustain themselves. We could even say that stars are self-cannibalizing entities that "eat" their own entrails in order to survive. Stars exist due to a constant tug of war between gravity trying to squeeze them inward and nuclear fusion trying to blow them apart. Amazingly, this seemingly unstable situation can render them stable for billions of years. Our sun, a modest size star a little under 5 billion years of age, is at roughly the middle of its "life cycle" (note the nomenclature). Stars form in regions rich in gases and chemical elements called "stellar nurseries." When a star exhausts the fuel at its core, it "dies" in a huge explosion, creating shock waves that propagate and spread their material through interstellar space. When these shock waves from the dying star collide with gas clouds, they may trigger the formation of new stars. In a loose sense, then, stars are reproducing, even sharing some of their original matter with the nascent ones.

There is poetry in the life-and-death cycle of stars, made meaningful due to our own experience of life. We are so imbued with life that we tend to see it everywhere. Still, stars are not alive in the way organisms are. They don't have a metabolism, and they don't reproduce to form historical lineages of open-ended length.

Organisms are fundamentally different from fires, hurricanes, and stars. They are not just self-organizing and self-maintaining. They're also self-producing and world constructing, and they generate historical lineages by reproducing and evolving. These are the things we are going to focus on.

The Disappearance and Return of the Organism

When we use the word *organism*, we mean a living being understood as an organized system. Indeed, the term *organism* was introduced in the eighteenth century to mean the kind of organization exemplified by living beings in contrast to the organization displayed by machines. Only later, in the nineteenth century, did the word come to refer to individual living beings.[5] Thus, the concept of the organism implicitly contains the idea that living beings have a unique organization, making them different from both nonliving physical things and machines.

It was precisely this idea that disappeared in the mid-twentieth century with the rise of molecular biology and the conception of living things as molecular machines. Molecular biologists, using methods from biophysics and biochemistry and explanatory concepts and models from cybernetics, information theory, and computer science, sought to explain all biological phenomena ultimately in terms of the structural properties of macromolecules. Earlier, in the 1930s, the architects of the so-called modern synthesis had merged Darwin's theory of evolution by natural selection with Mendelian genetics, recast as population genetics. With the subsequent identification of genes with segments of DNA, the organism, as an explanatory concept, disappeared in favor of molecular entities, on one side, and populations described in terms of their genetic composition, on the other. Although there were always dissenters from this framework, such as the Theoretical Biology Club in the 1930s (whose members were influenced by Whitehead's process philosophy of the organism) and Barry Commoner in the 1960s (later known for his environmental activism), the concept of the organism returned to prominence only toward the end of the twentieth century, spurred by forceful critiques of reductionistic molecular biology and the modern synthesis, and by a renewed interest in the theoretical problem of explaining the organization of life.[6]

Genocentrism, the position that the fundamental units of life are genes, not organisms, was the principal target of these critiques. According to genocentrism, organisms are vehicles made by and for genes. Although genes, strictly speaking, were supposed to be abstract functional entities obeying Mendel's laws of inheritance, they were now identified with lengths of DNA. This conflation of function and physical structure led to genes coming to serve the same reductionist purpose in biology as atoms traditionally did in physics.[7] Thus, during the latter half of the twentieth century, many molecular biologists understood their project to be that of explaining living phenomena as epiphenomena of genes.

But not all biologists thought this way. Already in 1965 Barry Commoner wrote, "The biochemical specificity that governs the precision with which inherited effects are transmitted from parent to offspring originates in no one molecule but in a complex network of molecular interactions. . . . The entire complex system of the living cell, including DNA and various enzymes, is the source of the biochemical specificity of the cell which is transmissible in inheritance. Despite modern 'molecular genetics,' the cell theory remains in force."[8] In other words, everything pertaining to bio-chemical specificity—DNA replication, protein synthesis, and metabolism (rates of reaction, linked series of chemical reactions)—happens inside a cell and by virtue of the enabling context and constraints of the functional organization of the cell as a whole. Thus, the cell, as a unicellular organism or as part of a multicellular organism, remains paramount. As Commoner wrote, "DNA is neither a self-sufficient genetic 'code' nor the 'master chem-ical' of the cell. . . . [I]nstead . . . inheritance depends on multimolecular interactions which occur in no system less complex than that in an intact cell; self-duplication is, therefore, not a molecular attribute of DNA, but a property of the whole cell."[9]

Nevertheless, the reductionist strategy of breaking up the organism (the cell) into its parts proved to be such a powerful method for investigating biomolecules that a crucial truth was forgotten: the properties of each part depend on the interrelations of the parts and on the context of the parts within the whole. As a result of neglecting this truth, an unwarranted ontology—life as a molecular machine—came to be surreptitiously substi-tuted for a useful method of investigation.

One prevalent effect of this molecular-biology case of surreptitious sub-stitution is "molecular animism," the projection onto molecules of the ani-mate agency proper to organisms. In philosopher Daniel Nicholson's words,

> This is the tendency to anthropomorphize molecules by calling them "regula-tors," "integrators," "organizers," etc., and crediting them with the regulatory, integrative, and organizing effects that actually arise from the coordinated activ-ity of the whole system. . . . The mistaken habit of bestowing privileged causal roles to molecules in explanations of cellular phenomena stems from the failure to understand that everything that happens inside a cell (e.g., DNA replication, protein synthesis, membrane trafficking, etc.) happens by virtue of the enabling conditions afforded by the pre-existing functional organization of the cell as a whole.[10]

The cell creates the stable, protective environment outside of which DNA replication and protein synthesis cannot happen, while the cell results from those molecular processes.

The growing recognition that understanding the whole in terms of its parts requires understanding the parts in terms of the whole lies behind the return of the organism as a fundamental explanatory concept for understanding biological complexity. Local molecular processes enabling development and evolution must be understood within the globally organized context of the cell that makes them possible in the first place.

But what exactly is this global organization? Clearly, it's not the sum of the parts, because the parts acquire their status as parts only by virtue of how they're organized in relation to one another. So, we need an independent account of the organization proper to the organism, particularly one that describes how the organism maintains its organization by continuously producing, breaking down, and replacing the parts that constitute that organization. This self-producing organization of the organism is our next topic.

What Makes Life Different: Autonomy and Agency

Ever since Erwin Schrödinger popularized the idea in his landmark book *What Is Life?*, first published in 1944, scientists have searched for universal principles that make living systems different from nonliving ones.[11] The hope has been to find a "grand unified theory" of life, ideally stateable in the language of physics. Nevertheless, seventy years after the appearance of *What Is Life?* no such theory has been found, despite many advances in theoretical and mathematical biology. There are different accounts of why this remains the case. Schrödinger speculated that for the structure of living matter, "We must be prepared to find a new type of physical law prevailing in it."[12] A few decades later, mathematical physicist Eugene Wigner suggested that present-day microscopic physics, particularly quantum physics, may presuppose situations in which life is absent and may be valid only under those special conditions, not universally. He speculated that the laws of physics may need to be modified to account for life.[13] Other scientists, however, whom we discuss below, argue that the key to understanding life is to introduce new principles of biological organization that do not contradict existing physicochemical laws but are not reducible to or expressible in terms of them.[14] One thing that's clear is that life's key characteristics put it in a separate category from just that of nonequilibrium dissipative structures.

Many of these characteristics revolve around the prefix "self." To be alive is to be self-individuating, to cleave existence into the dyad of self-versus-world. A living system constantly constructs itself, thereby making itself distinct from its environment while altering its environment in ways conducive to its own continuation. In other words, a living system possesses autonomy and agency: it produces, maintains, and regenerates the parts and processes that constitute its functioning as an integrated, self-governing whole (autonomy), while also both promoting environmental conditions favorable to its own existence and actively avoiding conditions that threaten it (agency). Autonomy and agency are what we need to investigate to understand how life differs from just dissipative structures.

Organization is the key concept for getting a handle on biological autonomy. Introduced by Immanuel Kant to describe the unique character of life, "organization" refers to how the parts of an organism exist for and by means of the whole and the whole exists for and by means of the parts.[15] The different parts of an organism play unique causal roles, while interrelating so as to generate and embody an integrated whole. According to Kant, organisms self-organize: their parts produce each other reciprocally and thereby constitute an organized whole that is also a condition for their own existence and operation.[16] To use an example unknown to Kant, enzymes in living cells are produced in metabolic pathways (linked series of chemical reactions) facilitated by other enzymes, which in turn are produced with the help of other enzymes, in a cyclic and recursive way, with all the metabolic pathways constrained by the boundary condition of the cell membrane, which itself is a product of metabolic pathways. This kind of self-organization is qualitatively different from the formation of physicochemical dissipative structures: it involves a huge diversity of molecular interactions that mutually enable and produce each other under the boundary condition of a membrane that results from a subset of those very interactions.

Kant's idea of biological self-organization was central to nineteenth-century biology, especially in continental Europe, and to embryology in the first half of the twentieth century. In the second half of the twentieth century, however, the idea was pushed aside by the rise of molecular biology and genetics, particularly by the misguided idea that genes determine biological organization. Although this idea initially shaped the outlook of molecular biology, it's logically detachable from it. Indeed, molecular biology itself makes clear that gene expression always occurs within a biologically self-organized system (a cell) and presupposes that system's presence as a necessary condition for its occurrence. Although genes play a crucial role in bringing about biological organization, they're also consequences of organization, not determinants of it.

In recent decades, the idea of organization has been retrieved and put back to work in biology.[17] A crucial step forward happened in the 1960s and 1970s with the idea that biological organization can be described using the concept of closure.[18] In its mathematical sense, "closure" means that an operation on members of a set always produces a member of the same set. For example, the set of real numbers is closed under addition because the addition of two real numbers always produces another real number. In the biological context, "closure" means that a process (such as protein synthesis) produces elements (such as enzymes) that make up and enable that process, so that the process and elements are mutually dependent.

Psychologist Jean Piaget was one the first theorists to apply the concept of closure to biological processes in his 1967 work *Biology and Knowledge*.[19] He described how an organism must be both thermodynamically open to exchanges with the environment and organizationally closed in the sense of being made up of elements that maintain a cyclic organization by regenerating themselves through chains of mutual dependence.[20]

A decade later, neurobiologist Francisco Varela introduced the concept of "organizational closure" and used it to define an autonomous system.[21] For a system to be autonomous, its constituent processes must have organizational closure. This means that the processes recursively depend on each other for their mutual generation as a network and thereby constitute the system as a unity. The paradigm is a living cell. In Varela's words, "A cell stands out of a molecular soup by defining and specifying boundaries that set it apart from what it is not. However, this specification of boundaries is done through molecular productions made possible through the boundaries themselves."[22] In the single-cell version of biological autonomy, the constituent processes are biochemical, and their recursive interdependence takes the form of a self-producing network that also generates and maintains its own boundary conditions, particularly its membrane. The creation of the membrane defines the inside and the outside, but only by having an inside and an outside can you have the processes that create the membrane. Varela, together with his mentor and colleague Humberto Maturana, called this kind of cellular autonomy "autopoiesis."[23]

Theoretical biologist Robert Rosen was another important scientist who used the concept of closure to describe the organization of the organism.[24] He used Aristotle's distinction between material cause (that out of which something is made) and efficient cause (the primary source of change) to state that an organism is "closed to efficient causation." This means that anything within an organism that is an efficient cause of some process is materially produced within the organism. In concrete terms, every catalyst (enzyme) required for metabolism is a product of metabolism, so no outside catalytic activity is necessary to maintain the organism.

Around the same time, Stuart Kauffman proposed and explored models of what he called "collectively autocatalytic sets." A collectively autocatalytic set of molecules is one that, as a whole, catalyzes the formation of all members of the set, with no molecule catalyzing its own formation. The collection thereby achieves "catalytic closure."[25]

In later work, Kauffman took the further step of thermodynamically grounding these ideas in what he called a "work-constraint cycle."[26] To make use of a flow of energy to generate work rather than heat, the energy flow needs to be constrained: its degrees of freedom need to be limited. Thus, work can be thought of as a constrained release of energy. If a system uses work to regenerate at least some of the constraints that make work possible, then it embodies a work-constraint cycle. In Kauffman's words, "work begets constraints begets work." This is the case in an organism. Unlike the walls of an engine cylinder, which constrain combustion, the constraints that enable an organism to channel the flow of energy are not preestablished and manufactured from the outside. Rather, they are produced and maintained by the organism itself. The organism uses the work generated by the constraints (such as membrane structure or an enzyme facilitating a chemical reaction) in order to generate those very constraints (to rebuild the membrane or to produce enzymes). Thus, biological autonomy, the fact that organisms, in Kauffman's words, "act on their own behalf," is thermodynamically grounded on the organizational closure of work-constraint cycles.

Theoretical biologists and philosophers Maël Montévil and Matteo Mossio, along with mathematician Giuseppe Longo and philosopher Alvaro Moreno, have synthesized and developed these ideas in their account of "closure of constraints" as the unique basis of biological organization and autonomy.[27] They understand a constraint as a structure that, while acting on a given process, remains unaffected by it at a specific timescale. For example, the vascular system transports blood without being altered by blood flow at the same timescale, and an enzyme catalyzes a reaction without being used up in it. In physics, constraints are typically independent structures that don't depend on what they constrain for their existence and maintenance. For example, an inclined plane constrains how a ball rolls, but the ball's rolling does not contribute to the existence and maintenance of the plane. In biology, however, constraints (the vascular system, metabolic pathways) typically do depend on what they constrain (oxygen delivery, enzymes) for their existence and maintenance. In particular, in an organism, "the existence of a set of constraints collectively depends on the actions that they exert on the processes and dynamics. When this occurs, the set of mutually dependent constraints can be said to realize closure and therefore to be organized."[28] Organizational closure, understood as closure of constraints, "achieves a form of 'self-determination,' in the precise sense that the conditions of existence of the constraints subject to closure . . . are determined within the organization itself."[29] Thus, closure of constraints entails autonomy (self-determination).

Autonomy through constraint closure underwrites biological function. The function of the heart is to pump blood, not to make sounds, because pumping blood contributes to the organism's constraint closure (autonomy), whereas making sounds does not.[30] In other words, out of the myriad causal effects of a given process, the effects required for constraint closure are the functional ones.

Autonomy (self-determination) also underwrites agency (being able to interact with the world so as to promote favorable conditions and avoid dangerous ones). Consider bacteria, among the oldest, smallest, and structurally simplest organisms. They are autopoietic systems that adaptively regulate themselves in relation to each other and changing environmental conditions. Bacteria communicate with each other through chemical signals and thereby regulate one another's gene expression in response to fluctuations in their cell population density, a behavior called "quorum sensing." They also move toward what they find attractive and away from what they find repellent, a behavior known as chemotaxis. Their movement isn't passively undergone; it's actively directed. Motile bacteria swim by means of flagella embedded in their membranes. They rotate their flagella to form a propeller that pushes the cell body forward or to tumble about randomly. As they move, they're able to register changes over time in the concentration of attractants or repellents. The cells maintain their heading as long as they detect an increase in the nutrient level. If the nutrient level decreases, the cells go into a random tumbling mode until they hit on an orientation where they again detect an increase, at which point they head in that direction. By repeating these behaviors—swimming in the same direction as long as conditions are improving or not getting any worse and tumbling when conditions start deteriorating or reversing course when they encounter toxins—bacteria can travel long distances toward favorable locales and away from deleterious ones. Quorum sensing and chemotaxis are examples of agency: the cells actively respond to their environment and change their relationship to it so as to maintain their autonomy.[31]

From this perspective, agency is the capacity of an autonomous system to regulate its interactions with the environment, and thereby to ensure its own maintenance and self-determination.[32] Adaptive agents modulate their behavior in relation to conditions registered as improving or deteriorating, viable or unviable, so as to preserve their viability.[33] They actively respond to their environments and each other. These capacities are the core characteristics of organisms. Unlike other physical systems, such as fires, hurricanes, and stars, organisms are autonomous and adaptive agents.

Living Is Sensemaking in Precarious Conditions

Autonomy and agency imply sensemaking. Organisms are sensemaking beings. They create worlds of relevance. They engage with environments that are structured by their actions and by what is relevant to their viability. Living beings partition their environments into things that have positive or negative value, or that are neutral, or have no significance at all, that fall outside the sphere of significance altogether. In Varela's words, living is sensemaking.[34]

Consider, again, bacterial chemotaxis. Bacteria swim up a gradient of sucrose, which they can metabolize as a nutrient. Sucrose means food for these organisms. Where does this kind of meaning come from? Looked at from a purely physicochemical perspective, the sucrose gradient is just a variable concentration of disaccharides (two-part molecules). It has no inherent meaning, significance, or value. Its being food is not intrinsic to its physicochemical structure. Nor is its being food just a matter of its being able to bond to other molecules in the cell membrane. Rather, sucrose means food only given the bacterium as an autonomous agent that must maintain its viability. Meaning resides not at the molecular level but at the level of autonomous agency, which is to say at the level of the organism as a whole. Sucrose means food because it contributes to the bacterium's ongoing self-determination (organizational closure). Sucrose has significance and value as food in the microbial world that emerges with bacteria and other microbes. Of course, this kind of significance depends on the physicochemical structure of sucrose—on its being able to form a gradient, traverse a cell membrane, and so on. Nevertheless, the significance of sucrose as food emerges only given the bacterial cell as an autonomous agent that must preserve its self-determination and stay within its zone of viability.

Some researchers use the concept of semantic information to describe this kind of significance.[35] The important distinction is between syntactic information and semantic information. Information theory, the theory of the communication of digital information originally formulated by Claude Shannon, is not a theory of semantic information (meaning, sense or significance). Instead, it provides measures of statistical correlations between events or systems. These are measures of so-called syntactic information. When we say, "smoke means fire" or "tree rings indicate age," we are using "means" or "indicate" to stand for physical correlations. But when we say "sucrose means food for a bacterium," "means" cannot be reduced to statistical correlations of sucrose, membrane transduction, and metabolic uptake. "Semantic information" refers to correlations that carry significance for a system's self-maintenance as an autonomous agent. Sucrose carries significance (semantic information) for the bacterium as a self-determining (autonomous) agent that must maintain its viability. The bacterium registers conditions as improving or deteriorating and works to keep itself within its zone of viability. The sucrose gradient is a difference that makes a difference to its viability. The link between sucrose and viability (between swimming up a sucrose gradient and increasing viability) has significance (carries semantic information) for the bacterium as an autonomous agent.

The need to maintain viability arises from its constantly being challenged in the precarious conditions of a thermodynamic universe. Thus, Varela's statement that living is sensemaking can be extended by saying that *living is sensemaking in precarious conditions*.[36]

Imagine that you are very small, so that you are continually buffeted by water molecules and bumped off course, while the watery contents inside you are in constant motion. Such is the external and internal milieu of bacteria, the microworld of thermal diffusion and Brownian motion (the random motion of particles suspended in a liquid or a gas). How do you hold together as a living being? You depend completely, of course, on the physicochemical properties of strong and weak bonds, but you also hold together because you are an autonomous agent with organizational closure: every one of your constituent processes is both enabled by and is an enabling condition for one or more of your other constituent processes. Precarious conditions are ones in which the constituent processes (catalysis, enzyme regulation, membrane formation and repair) cannot sustain themselves in the absence of organizational closure in otherwise equivalent physical situations.[37] More simply stated, these processes cannot last outside the protective structure of the cell. Remove them, and they will tend to run down or atrophy. All life is precarious in this sense. Break open a cell, and its metabolic constituents diffuse back into a molecular soup; take an ant out of a colony, and it eventually dies; remove a person from a relationship, and they may cease to flourish. Precarious conditions entail the constant need for adaptivity, for regulating one's activity and behavior in conditions registered as advantageous or deleterious with respect to one's viability in a constantly changing and nonstationary environment (one whose statistical properties change over time). As Ezequiel Di Paolo says, "Life would not be better off without precariousness; it simply would not be life at all."[38]

What Makes Life Different: Open-Ended Evolution

"Nothing in Biology Makes Sense Except in the Light of Evolution" is the title of a 1973 essay by renowned evolutionary biologist and geneticist Theodosius Dobzhansky.[39] His statement, now famous, is often repeated in biology textbooks. (Long forgotten is Dobzhansky's having made this statement in the context of rejecting creationism while arguing for the compatibility of belief in God and the fact of evolution.) But the statement needs qualification given the foregoing ideas about biological autonomy.[40] Some things in biology do make sense apart from the light of evolution. In particular, biological autonomy, defined as closure of constraints, and agency, acting on one's behalf, are organizational concepts, not evolutionary ones, even though all existing autonomous agents are products of evolution. The point is conceptual: organizational concepts aren't reducible to genealogical ones, and evolution cannot happen without organized systems that reproduce with variation. So, the evolutionary perspective on life needs to be combined with the organizational one.

Evolution, in Darwin's words, is descent with modification. It results from variation, inheritance, and natural selection, as Darwin showed. But it also requires autonomous (organizationally closed) systems that reproduce. Biological evolution could not have gotten underway without minimal autonomous systems, such as autopoietic protocells, already in place. A minimal degree of organized complexity is required for complexity to increase as a result of natural selection.[41] Thus, organization is a condition for evolution. But evolution is required to go from minimal autonomous systems in the form of autopoietic protocells to full-fledged autonomous agents in the form of organisms. Descent with modification and natural selection have led to cascades of increased organizational complexity over the eons of life's history on Earth. Organization as we know it today, and as we're able to reconstruct its history, is a result of evolution. Life presumably started with simple autopoietic protocells (systems with catalytic-constraint closure that produced their own membrane), effecting the transition from an abiotic to biotic world, followed by the evolution of prokaryotes (archaea and bacteria), single-cell eukaryotes, and multicellular organisms (animals, plants, and fungi), down to today's biosphere (see chapter 9).[42]

Open-ended evolution is another characteristic that makes life different from other physical systems. "Open-ended" means that evolution has the capacity to generate an unlimited variety of living systems with no predetermined limit on how they physically realize organizational closure.[43] Furthermore, open-ended evolution generates not just an unlimited variety of organisms. It also generates evolvability, the capacity for adaptive evolution in populations of organisms. Evolvability too can increase in complexity as it evolves.

All of this happens at a much longer timescale than that of the lifetimes of individual organisms. It also happens at a collective and planetary scale. Populations of organisms create new worlds for each other, especially through symbiosis, and they cycle biological, chemical, and geological elements at a global scale (see chapter 9). Life in the fullest sense is a historical, collective, and planetary phenomenon. If we define living beings as autonomous (organizationally closed) agents with capacities for open-ended evolution, we must also say that any living being, or population of living beings, necessarily is enmeshed in the historical, collective, and planetary web of life.[44] Life depends on autopoiesis, evolution, and ecopoiesis—on the self-production of the cell as the basic unit of life; on descent with modification and natural selection, leading to the vast proliferation of diverse organisms; and on the collective, global production, and self-maintenance of the biosphere.

Organisms Are Not Machines

Organisms, understood as organized, autonomous, sensemaking agents capable of reproduction and open-ended evolution, are fundamentally different from machines.[45] In general terms, a machine is a device consisting of an assembly of components that operate in a coordinated way according to a prescribed sequence so as to produce a prespecified outcome. The properties of the components are independent of the whole, and they exist prior to the whole. Furthermore, the purpose or function of the machine is extrinsic: it lies outside the machine in its maker or user. In an organism, however, the properties of the parts (the functional components) depend on the whole, and the parts do not exist before the whole: components, such as enzymes or organelles, originate from previous cells and are synthesized or produced within the cell. In addition, an organism, as an autonomous agent, acts on its own behalf, so it is intrinsically purposive: it acts toward its own ends, such as maintaining its viability in precarious conditions. Thus, intrinsic purposiveness derives directly from biological autonomy and agency.

Machines are tailor made for reductive explanations, ones that explain a system by breaking it down into its separable components. This is because machines are "decomposable."[46] In a decomposable system, the intrinsic (nonrelational) properties of each part determine its operation independent of the other parts. If the system is arranged hierarchically, so that parts within a subpart interact more than do parts belonging to different subparts, the system is said to be "nearly decomposable." Hierarchical or nearly decomposable systems are also well suited to reductive explanation. By contrast, in a "nondecomposable" system, how the parts behave depends strongly on their dynamic interrelations and the changing context of the whole. In this case, machine models and reductive explanations have limited value.

The irony is that recent molecular-biology investigations inspired by the machine model of life have produced experimental data that undermine that model and reveal the limitations of molecular reductionism.

Nicholson reviews four research areas where this has happened: cellular architecture, protein complexes, intracellular transport, and cellular behavior.[47] According to the machine conception of the cell, cellular architecture is a static, highly ordered structure; protein complexes are specialized, molecular machines; intracellular transport is the movement of substances by miniature engines propelled by mechanical forces; and cellular behavior (such as gene expression) occurs as the result of a deterministic program encoded in the genome. New findings, however, challenge this picture and suggest an alternative view: "cellular architecture is regarded as a fluid, self-organizing process; protein complexes are considered to be transient, pleomorphic ensembles; intracellular transport is deemed to result from the harnessing of Brownian motion; and cellular behavior is viewed as a probabilistic affair, subject to constant stochastic fluctuations."[48] In other words, a cell continuously transforms its internal architecture by balancing molecular flows that are far from equilibrium; proteins fold into many different shapes (conformations) depending on the context; substances move around directionally within the cell by surfing the storm of random fluid motion engulfing them; and cellular behavior does not unfold according to a step-by-step algorithm, but rather involves random, context-dependent events in an environment subject to the chaotic dynamics of Brownian motion. Nicholson concludes his review and analysis of these findings as follows:

> The cell is *not* a machine, but something altogether different—something more interesting yet also more unruly. It is a bounded, self-maintaining, steady-state organization of interconnected and interdependent processes; an integrated, dynamically stable, multi-scale system of conjugated fluxes collectively displaced from thermodynamic equilibrium. Given its precarious nature, the cell is constantly having to negotiate a trade-off between structural stability and functional flexibility: too much rigidity compromises physiological adaptability, and too much promiscuity compromises metabolic efficiency. The cell accomplishes this by continuously turning over and reorganizing its constituents into different macromolecular complexes with diverse functional capabilities, which assemble and disassemble in order to meet the ever-changing demands of the environment. The permanent stochastic shuffling of molecules inside the cell and their opportunistic associations to form transient functional ensembles in response to intracellular and extracellular cues provides fast and robust solutions to the adaptive problems faced by the cell in a way that strikes an optimal balance between efficacy and plasticity.[49]

Nicholson emphasizes that every cell, whether in an organism or as a unicellular organism, is structurally and behaviorally unique. No two cells or organisms, even if they are genetically identical (isogenic), behave in exactly the same way. Cellular behavior incorporates an essential probabilistic component, and cell populations are heterogeneous. Traditional methods that average across populations lose sight of the uniqueness of the cell as an individual and the variability in the cellular world. "Looking to the future, as cell biology progressively morphs into 'single-cell biology' and we devote increasing attention to carefully characterizing not just individual cells, but also individual molecules in individual cells, we may soon find ourselves in the position of having to reconsider our understanding of even the most basic biological processes."[50]

Indeed, such reconsideration is already well underway, as biologists, mathematicians, and philosophers work together to understand life as "based on physics but beyond physics," to use Stuart Kauffman's words.[51] To see what this means, we need to return to the idea of phase space, which we introduced in chapter 1 as a key element of the ascending spiral of abstraction in mathematical physics and took up again in chapters 3 and 5 in our treatments of time and cosmology.

Phase Space Revisited

Guiseppe Longo, Maël Montévil, and Stuart Kauffman have argued that it is impossible to "prestate" the phase space of biological evolution.[52] Evolution is constantly changing in ways that make it impossible to specify in advance the observables (measurable physical quantities), parameters (degrees of freedom), and boundary conditions (initial conditions and constraints) of evolution's space of possible trajectories (phylogenetic pathways). It follows, they say, that "the strong reductionist dream of a theory that entails the full becoming of the universe is wrong. With life, we reach the end of a physics worldview that has dominated us since Newton."[53] In our terms, with life, we reach the end of the Blind Spot worldview.

Recall that phase space is an abstract space in which all possible states and trajectories of a system are represented, with each possible state corresponding to a unique point in the space. In classical mechanics, phase space consists of all possible values of position and momentum variables.

As the system evolves in time, it traces out a trajectory in phase space, and the shape of the trajectory reveals aspects of its dynamics.

The phase space of a system must be constructed in advance of calculating the trajectories in it. This is what Longo, Montévil, and Kauffman mean by the space being "prestated." In billiard ball physics, for example, the phase space is constructed by specifying the boundary conditions (the surfaces of the billiard table), the initial conditions (the starting position of the balls), and all possible positions and momenta of the balls. Using Newton's three laws of motion expressed as differential equations, one solves for the trajectories of the balls by integrating the equations, given the initial and boundary conditions. In other words, one deduces the trajectories by deriving the solutions of the equations. Thus, the trajectories in phase space are entailed (logically necessary), given the laws and the initial and boundary conditions.

Such entailment holds even for systems we can't predict due to what is known as deterministic chaos. Since Poincaré, we have known that prediction is impossible for certain kinds of systems, such as two or more planets and the sun, or the many billiard balls (the so-called three-body or n-body problem). These chaotic systems are nonlinear (effects are not proportional to causes), their behavior is aperiodic, and their trajectories are extremely sensitive to tiny changes in the initial conditions. Nevertheless, they are deterministic: their behavior is fully determined by the initial conditions and equations of motion. So although these systems cannot be predicted (their equations of motion cannot be analytically solved), their trajectories in phase space remain entailed by the equations, given the initial and boundary conditions.

We should recall that phase spaces are not "out there," hidden behind phenomena, waiting for us to discover them. They are abstract constructions. Physicists arrived at the concept of phase space over centuries by analyzing trajectories (Galileo), placing them in an analytic geometrical space (Descartes), specifying laws of motion (Newton), defining observables (such as momentum), and progressively elaborating structural invariants (constraints, generalized coordinates). We described in chapter 2 this ascending spiral of abstraction due to advances by Lagrange, Hamilton, Poincaré, Boltzmann, and many others.[54] But as Longo, Montévil, and Kauffman write, "Physical (phase) spaces, thus, are not 'already there,' as absolutes underlying phenomena: they are our remarkable and very effective invention in order to make physical phenomena intelligible."[55]

The crucial point, they go on to say, is that this method hits a hard limit in the case of life. Phase spaces for physical systems are maps of the known possible. Billiard balls and planets can go only certain ways within phase space. Evolutionary pathways, however, are not confined in this way, because the phase space of evolution is constantly changing. Imagine the surface of a billiard table constantly undergoing deformation, including adding new holes in the table, partly as a function of the balls constantly reproducing and changing in size and texture. Evolution is even more complex, given the diversity of species and niches. Organisms and environments co-create each other in a circular way. The parameters and boundary conditions of phylogeny constantly change at molecular, morphological, behavioral, and ecological levels. On the one hand, the parts of an organism can take on new functions that are not determined in advance: feathers originally helped to regulate temperature, but then came to be used for flight. Stephen Jay Gould and Elisabeth Vrba called this co-opting of a phenotypic feature by selection for a new function "exaptation."[56] On the other hand, evolution creates new niches. As phenotypic functions change, new worlds for organisms are opened up, and these worlds make possible new organisms, which in turn create new worlds. None of this is determined in advance according to any biological "laws of motion." The evolving biosphere is not "entailed" by laws, but instead is "enabled" by history.[57] Evolution occurs through the proliferation of contingent pathways and the exploration of what Kauffman calls the "adjacent possible" (the possibilities available to phylogeny at a given time). There is no way to determine in advance the phenotypes and niches that will co-create each other in the course of evolution. We can retrospectively explain such co-creation after the fact, but we cannot predefine phase spaces in which these co-created pathways are possible trajectories. Kauffman writes: "Because we cannot prestate the new functions that emerge and form the ever-changing phase space of the evolving biosphere, we can write no laws of motion for the evolving biosphere. We cannot integrate the equations we do not have. Therefore, no Newton-like laws entail the evolution of our or any biosphere among the 10 to 22 estimated solar systems in the universe."[58] This is what it means to say that life has no prestateable phase space.

The implications of this perspective are decisive for the Blind Spot, particularly for physicalism and reductionism. According to these ideas, the fundamental laws of the physical universe govern the spatiotemporal arrangements of matter and energy at the lowest levels, and everything else is determined by those laws and arrangements. But life—to say nothing of experience—already escapes this framework. If the evolving biosphere involves an ever-changing phase space that continually opens up new "adjacent possible" pathways, then life generates new forms and possibilities in ways that the metaphysics of physicalism and reductionism cannot capture.[59] As Kauffman puts it, "In short, life is based on physics but beyond physics. There can be no 'final theory' for the evolution of a universe having at least one evolving biosphere."[60]

Life beyond the Blind Spot

Francisco Varela was fond of quoting the poet Antonio Machado: "Wanderer the road is your footsteps, nothing else; you lay down a path in walking."[61] Life lays down paths in walking, none of them predefined and entailed by antecedent conditions. If life is based on physics but beyond physics, then we're going to need new scientific ideas for thinking about how life lays down paths in walking.

Biologists, together with physicists, mathematicians, and philosophers, are busily developing these new ideas.[62] Our purpose here is not to review them, since they are still in their early stages. Instead, we emphasize that biology has already brought us to the point where we can glimpse a view of a science of life beyond the Blind Spot. Organisms are autonomous agents, not machines. Life comprises unprestateable phase spaces. Life is self-enabling rather than law entailed. Life is essentially historical.

Physicalism, reductionism, and epiphenomenalism—keystones of the Blind Spot worldview—all fail for life.

We end with Kauffman's words: "This vast emergent becoming is beyond physics, yet based on it. This is life co-constructing itself and enabling its own vast evolutionary diversification here, and on any biosphere, in the universe."[63]

7 Cognition

It's Complicated

The twentieth century gave birth to a new scientific project, the transdisciplinary science of the mind called cognitive science, which combines anthropology, artificial intelligence (AI), linguistics, neuroscience, philosophy, and psychology. Its relationship to the Blind Spot is complicated. On the one hand, cognitive science expands the Blind Spot into the field of the mind; on the other hand, it offers a unique opportunity to reveal and try to move beyond the Blind Spot. The history of cognitive science, particularly the central role AI has played in it, makes this ambiguity evident.

When cognitive science arose in the 1950s, it was based on the idea that the mind is essentially a computer. The mind is software; the brain is hardware. Cognition consists of computations performed by the brain or an artificial system (a computer or robot). Cognitive scientists originally considered these computations to consist of symbols having the form of words or phrases. Cognition was thought to be the manipulation of language-like symbols in the brain according to logical rules. In subsequent decades, this idea, known as the classical computational theory of mind, came under criticism from a variety of perspectives. Neural network theory (also called connectionism) proposed that the computations underlying intelligent behavior are numerical in form rather than linguistic and logical. Cognitive processes consist of "subsymbolic" activation patterns in biological or artificial neural networks. This framework is the basis of today's field of machine learning, in which computers learn to perform tasks (such as categorizing objects in images) without being explicitly programmed to do them. A related approach came from dynamical cognitive science. It uses the tools of dynamical systems theory from mathematics and physics to model how cognitive processes and intelligent behavior unfold in time, either in neural networks or whole agents interacting with their environments, in contrast to classical computational models, which neglect the timescale over which mental computation occurs. Finally, the embodied cognition research program, particularly the version known as enactive cognitive science, emphasized the importance of the body and interactions with the environment for understanding the mind. Cognition is not located in the head, but instead resides in the ongoing, active relation between the embodied agent and the physical and social environment. Cognition of a meaningful world is brought forth or enacted through situated bodily action.

These different perspectives make the relationship between cognitive science and the Blind Spot ambiguous. On the one hand, the computational theory of mind, especially its implementation in AI, enlarges the Blind Spot by surreptitiously substituting computation for meaning and the experience of understanding. On the other hand, cognitive science quickly finds itself needing to come to terms with the Blind Spot, because direct experience remains the indispensable touchstone for understanding the mind.

Cognitive science differs from physics and biology, the sciences we have considered so far, with regard to the Blind Spot. The human observer usually can be set aside in physics and biology or be redefined operationally in terms of some procedure for measuring things, such as measuring time by noting the numerals on a physical clock. Of course, this move to "bracket the observer" runs up against significant limits in the fundamental physics of time and matter, and the biology of life, as we discussed earlier. Still, cognitive science cannot get away with bracketing the observer so easily as these other sciences can. Cognitive science is the science of the mind, and observation is a mental capacity based on conscious perception. It follows that when our concern is the mind, observation cannot be set aside or redefined operationally without presupposing what needs explaining. So we must confront and take account of direct experience when doing cognitive science. The question is whether we suppress or marginalize experience, or whether we recognize the primacy of direct experience and accordingly try to reorient the science of the mind away from the Blind Spot. This question becomes most pressing in the cognitive neuroscience of consciousness, which we discuss in the next chapter.

Whereas classical cognitive science tried hard to ignore experience in favor of nonconscious computational processes, enactive cognitive science arose by recognizing embodied experience as the unavoidable reference point for understanding the mind.[1] Enactive-cognition theorists embrace the irreducible primacy of direct experience and accordingly strive to move beyond the Blind Spot in their investigations of the mind.[2] The tension between these divergent approaches is another sign of the unresolved ambiguity in cognitive science's relationship to the Blind Spot. We return to this ambiguity at the end of the chapter.

It is impossible to cover all the ways that the subdisciplines of cognitive science, individually and collectively, relate to the Blind Spot. Our focus instead is the strongest and most prominent manifestation of the Blind Spot in cognitive science, which we find in AI.

Relevance Realization

Thinking that the mind is essentially a computer has given rise to intractable puzzles, visible especially in AI. They center on meaning and understanding, particularly on how we understand relevance.[3]

Consider note taking during a lecture. Taking good notes requires zeroing in on what is relevant and knowing how to ignore what is irrelevant. You need to be able to follow a train of thought and extract its main ideas while making connections to other things you know. If you ask questions, whether to get more detail, clear up a misunderstanding, or bring out and explore presuppositions or implications of what is being said, you also have to be able to discern relevance. Finally, this whole activity is shaped and guided by norms, by rational standards of how you should think and speak, social standards of acceptable behavior, and commonsense standards of good judgment. Understanding and being able to follow these norms also is a crucial part of "relevance realization."[4]

Despite AI hype, which tends to wax and wane with AI's successes and failures over the decades, no AI system comes close to realizing relevance, and no historical or current AI approaches come anywhere near being able to create systems that understand meaning, have general intelligence, or manifest even minimal common sense.[5]

AI research has several distinct but interwoven threads. We can distinguish among AI as the science and engineering of making computational systems, AI as part of cognitive science, and the focused research effort sometimes called "strong AI."[6] Strong AI is the position that a computer with the right kind of program has a mind and that the human mind is a collection of programs running on the brain. Strong AI theorists seek to explain the mind by trying to build computational systems that implement or instantiate cognitive abilities, such as recognizing faces or translating languages, while looking to psychologists and neuroscientists to figure out which programs or algorithms are instantiated in the brain and underlie human cognition.

Although the field of machine learning has created systems that perform tasks such as playing games, detecting objects in images, and generating spoken or written text, the limitations of their performance, especially with respect to genuine cognition and understanding meaning, have been shown to be readily apparent with a little probing and analysis.[7] "Adversarial examples" are a well-known case. Many machine-learning image recognition algorithms can be tricked into making large classification errors, such as classifying images of buildings, vehicles, dogs, and insects as images of ostriches, by making trivial changes to the pixels that leave the images unchanged to humans.[8] Adversarial examples indicate that these systems learn statistical correlations, not recognitional concepts for objects, and that the systems have no understanding of objects as such. Moreover, no AI system has anything like broad, general-level intelligence—the ability not only to solve but also to pose new problems by integrating multiple cognitive abilities and drawing from multiple sources of knowledge—and achieving this kind of intelligence is nowhere in sight in AI. In Brian Cantwell Smith's terms, AI systems excel at "reckoning" (calculation) but are completely lacking in "judgment."[9] A key problem is getting AI systems to realize relevance, as the following examples will indicate.

The Frame Problem

AI's "frame problem" illustrates the importance of relevance and the difficulties with getting computational systems to realize relevance.[10]

Philosopher Daniel Dennett, in a classic article, illustrates the frame problem with a story about a robot trying to find batteries (its food source) and move them to a storage area.[11] The robot comes upon a wagon where there is a battery, but there is also a ticking time bomb on the wagon. The robot correctly deduces that if it pulls the wagon, the battery will come along with the wagon. But the robot fails to realize that the bomb also will come along with the wagon. The robot has failed to discern something crucially relevant to its actions. The robot's designers try to fix the problem by updating its program so that it deduces not just the intended effects of its actions but also the potential side effects. But now the robot just sits there, calculating the indefinitely large number of potential side effects, setting each one aside, until the bomb goes off. Again, the robot has failed to frame the situation correctly so that it considers only relevant information in a timely way. The designers decide to modify the program so that the robot will compute a list of the potentially relevant side effects it needs to consider. But which side effects are the relevant ones? This question reintroduces the original problem of how to determine which information is relevant versus which information is irrelevant. Furthermore, even if this problem can somehow be solved, how are the designers supposed to prevent the robot from once again stopping and calculating endlessly, this time by computing two lists, one with all the potentially relevant side effects and the other with all the irrelevant side effects?

The frame problem arises in trying to design an intelligent system (a computer program or robot) to act appropriately in real-world situations to accomplish some end. The system needs to be able to frame situations correctly so that it takes into account both intended and unintended side effects of its actions. It needs to focus on what is relevant and ignore what is irrelevant. How do you design or program an agent so that it considers only the relevant information without wasting its time on the irrelevant information?

Dennett's story illustrates that an agent must be able to ignore large amounts of information in an intelligent way to act in the world, even in relatively simple situations. An intelligent agent must be able to home in on what is relevant without even considering most of what is irrelevant. The paradox is that if the agent is able to ignore what is irrelevant, then it must somehow already grasp what is relevant.

The frame problem arose in the context of the classical computational theory of mind and first-wave, logic-based AI.[12] The problem illustrates the intractable difficulty of trying to specify relevance in classical (symbolic) computational terms—in propositional representations of situations in the world, along with heuristics (problem-solving shortcuts that find acceptable but imperfect solutions in a limited amount of time) for determining relevance. Human beings are able to deal effectively with open-ended and constantly changing situations in which anything could be relevant. Relevance is context dependent, and there is no limit on what can be relevant. It follows that relevance cannot be prespecified. But then how could it be captured in a set of language-like representations of the world together with a set of heuristic rules for manipulating them? Relevance realization, the evidence suggests, cannot be reduced to heuristic rules and symbolic representations.

The inability to realize relevance using language-like representations and heuristics is one of the motivations for second-wave, machine-learning-based AI. Here the aim is to design systems that can generate meaning and relevance for themselves out of huge amounts of data using very fast, parallel-processing hardware and learning algorithms. Although machine learning has been very successful within certain limited task domains, such as game playing, which we discuss next, present-day machine-learning systems lack many of the abilities crucial to relevance realization, such as being able to generalize across many different task domains and make analogies between one area of knowledge or experience and another one. Trying to take these systems out of their limited task domains and make them function autonomously in the everyday world immediately brings back the frame problem and the problem of relevance realization.

The Relevance of Games

Playing games like chess or Go is touted as an AI success story. But the way computers play games is very different from how we do, and the differences are revealing about relevance.[13]

AI treats game playing as a form of problem solving. In classical cognitive science, problem solving was described as moving through a search space from an initial state to a goal state through a sequence of operations that turn a given state into a subsequent state. To win a chess game, you have to go from the initial state of the opening position to the goal state of checkmating your opponent. But chess has a massive search space: the large number of possible moves and positions in the opening and middle game increases exponentially at each turn. The number of pathways is much too large for any computer, let alone a human being, to search exhaustively.

IBM's breakthrough chess-playing program, Deep Blue, which defeated world champion Garry Kasparov in 1997 using special-purpose hardware and brute-force computing power, could search only around twenty moves ahead and usually searched only to a depth of six to eight moves. Expert human chess players, however, use their pattern-recognition abilities for good chess positions and moves to rule out large regions of the search space as irrelevant and to focus only on the relevant regions. In other words, they rely on ways of zeroing in on relevance. In contrast, Deep Blue computed a partial branching tree of possible moves, using the current board position as the root, and then applied its evaluation function to estimate the values of positions. Although Deep Blue did extremely well at a game that requires intelligence when done by humans, its performance was based on brute-force computing power and a huge database of chess information spoon-fed to it by human players. (Deep Blue was not a learning system; its chess information was hand-programmed.) All Deep Blue could do was play chess, and its performance did not involve anything like human chess comprehension. Deep Blue certainly did not know it was playing chess or that it was in a contest with a human opponent.

Go has an even larger search space than chess. There are around 250 possible moves from any given board position in Go, whereas there are roughly 30 in chess. So using even limited forms of brute-force search is not workable. In addition, giving computer programs evaluation functions for Go board positions based on explicit, well-defined criteria does not lead to high-level play. Expert Go players say that they rely instead on their intuitions about whether board positions have "good shapes." Here relevance realization consists in being able to zero in on such shapes. Top players use their finely honed pattern-recognition skills, which they cannot verbalize or formulate in terms of explicit concepts and rules. For these reasons, Go presents a bigger challenge to AI than chess.

In 2016, AlphaGo, a Go-playing computer program created by Deep-Mind, defeated world champion Le Sedol. Unlike Deep Blue, AlphaGo was based on machine-learning algorithms in artificial neural networks. It used a class of algorithms called "deep learning" algorithms ("deep" refers to the number of layers in the neural network). AlphaGo worked by first learning to find patterns in games played by expert human players and to make the same moves experts make. Its designers then made it play repeatedly against earlier versions of itself to improve its performance (its chances of winning) using reinforcement learning (a machine-learning method where the program receives rewards for its actions and works to maximize cumulative rewards). The result was not only a computer program that could beat the best human Go players, but one that seemed to replicate their intuitive pattern-recognition skills. From the outside, it looked as if it was able to distinguish between relevant versus irrelevant board positions and moves better than its human opponents.

We're getting close to the Blind Spot. If AlphaGo, or its more powerful successor, AlphaGo Zero, "mastered" Go "without human knowledge" or "human guidance," as its inventors claimed, and if it "replicated" or "captured intuition," as some commentators claimed, then it is tempting to suppose that intuition and relevance realization are just a matter of computation.[14] The idea would be that what Go players, based on their subjective experience, call "intuition" consists of neural computations, which are subjectively inscrutable to the players themselves. Neural computations call the shots; they have causal power. The experience of intuition is an after-the-fact side effect, a subjective epiphenomenon of brain computations. According to this way of thinking, the apparently direct experience of intuition is part of the brain's user illusion.

We've now arrived at the Blind Spot. Although AlphaGo constitutes an AI milestone, and human intuition and relevance realization unquestionably depend on subjectively inscrutable brain processes, it is wrong to think that deep-learning methods obviate the need for artificial neural networks to rely on human knowledge, that these networks capture intuition, and that intuition is epiphenomenal. This line of thought perfectly illustrates the Blind Spot, and its every step is mistaken.

AlphaGo did not master Go "without human knowledge," despite the title of the paper in *Nature* by its inventors asserting the contrary.[15] As psychologist Gary Marcus and computer scientist Ernest Davis write, "The system still relied heavily on things that human researchers had discovered over the last few decades about how to get machines to play games like Go, most notably Monte Carlo Tree Search [a computer algorithm designed for game-playing programs] . . . which has nothing intrinsic to do with deep learning. Deep Mind also . . . built in the rules and some other detailed knowledge of the game. The claim that human knowledge wasn't involved simply wasn't factually accurate."[16] The inability to see that behind and within AlphaGo lie millennia of human experience playing Go, combined with decades of accumulated human knowledge about how to design game-playing programs, constitutes a striking case of the amnesia of experience at the heart of the Blind Spot.

AlphaGo does not replicate or capture human intuition, and it does not realize relevance, because it knows nothing about Go as a game. AlphaGo excels at detecting patterns that *we know are Go board positions*, but it does not know that its data structures represent Go positions and moves. AlphaGo does not know that it is evaluating Go moves; indeed, it does not know that it is playing Go at all. As Smith writes, "It is unlikely that AlphaGo and its successors have any sense of the fact that Go is a *game*, with an illustrious millennial history, played by experts from around the world—or even, for that matter, and more pointedly, that there is a difference between the particular game it is playing and the representation of that game in its data structures."[17] We understand AlphaGo's behavior as being about Go, a game that exists in the world outside its data structures. But AlphaGo has no knowledge of there being an outside world, let alone one with games. It has no way of understanding that what it is doing constitutes playing a game. Indeed, it has no understanding of anything; it is a sophisticated device for detecting complicated structural patterns and regularities in very large amounts of data, but it has no access to the semantics of the data, to what its data structures mean (to how they map onto something outside the system). AlphaGo does not replicate intuition, and deep-learning systems in general do not "bottle intuition."[18] Rather, AlphaGo's pattern-detection performance, *as interpreted by us from the outside*, models aspects of human Go-playing intuition by simulating them.

AlphaGo, and deep-learning artificial neural networks in general provide no evidence that intuition is epiphenomenal. On the contrary, the experience of intuition is how we should expect a highly skilled, embodied agent's perception to be organized so as to be efficacious for action in the world. Suppose we were to create an artificial, expert Go player that did know it was playing Go. To have this kind of knowledge it would have to know and understand numerous things about the world, including understanding itself as an agent in the Go-playing community and the world at large. We should expect that its deep, inner workings would be subjectively inscrutable to it precisely because it would be oriented to the world. We should expect that it would perceive Go board positions by intuiting them as having "good shapes" and that its intuitive mode of perception would be a causally efficacious way for its cognition to be organized in relation to the world. Without being organized for intuition, it would not be able to realize relevance, not just in the microworld of Go play but also in the world at large.

More Than a Game

Chess and Go are special. They have perfectly definite states, rules, moves, and pieces, and the players have access to information about these features at all times. Chess and Go constitute their own abstract, separate universes. Human beings can mentally inhabit both universes, understanding them not just as formal systems (as perfectly definite, abstract structures) but also as games having meaning. Humans can switch back and forth between playing each game, even at the same time. AlphaGo cannot do any of this. Computer scientist Melanie Mitchell writes that "even the most general version, AlphaZero, is not a single system that learned to play Go, chess, and *shogi*. Each game has its own separate convolutional neural network that must be trained from scratch for its particular game. Unlike humans, none of these programs can 'transfer' anything it has learned about one game to help it learn a different game."[19]

The human ability to generalize from one area of knowledge to another area, or to use an old skill to acquire a new one, involves being able to abstract and form analogies across very different domains of thought and action. In Mitchell's words, "You can show a deep neural network millions of pictures of bridges, for example, and it can probably recognize a new picture of a bridge over a river or something. But it can never abstract the notion of 'bridge' to, say, our concept of bridging the gender gap."[20] This ability to abstract and form analogies is crucial to relevance realization.[21] Although machine-learning research on "transfer learning" and open-ended learning in virtual agents and environments continues to advance, it is far from producing systems that can make meaningful analogies in the real world.[22]

Not all games are like chess and Go in being perfectly definite, self-contained abstract universes. The rules of charades, for example, vary widely from place to place, and charades does not divide up into discrete and definite states because it depends on continuous, improvised physical acting and mime. Playing charades is closer to real-world action. Mitchell writes (drawing from an article by Gary Marcus) that charades "requires sophisticated visual, linguistic, and social understanding far beyond the abilities of any current AI system. If you could build a robot that could play charades as well as, say, a six-year old child, then I think you could safely say that you had conquered several of the 'most challenging domains' for AI."[23]

Unlike a board game, the world does not come preconfigured into ready-made states governed by precise rules for how to act with a limited and definite set of elements. Take an everyday activity such as making dinner. Despite being a routine activity, making dinner cannot be delimited such that it consists of definite states and actions with a clear frame around them that determines relevance. Making dinner comprises a multitude of possible actions, objects, and states depending on the social setting, cultural norms, what you decide to make, where you are cooking, what is available to you, and how much time you have. Are you a sous chef in a fancy restaurant? Or a volunteer in a soup kitchen? Are you a competitor on a cooking show? Or making dinner for your family? Are you at home or camping? Here are some more possibilities: You don't have all the ingredients you want, so you have to improvise. You decide to try something new, so you consult internet recipes, which means inhibiting your impulse to check your email, social media, and the news. Your phone rings while you're at the stove and you have to decide whether to answer it. Your partner arrives, upset by something that happened at work, so you must divide your attention to listen to them while you're trying to cook. A power outage happens, so you have to replan everything on the fly. Or the fire alarm in your building goes off and you have to drop what you are doing and leave immediately. In real-world situations, there is no way to specify in advance what can happen and what may turn out to be relevant. There is no way to put boundaries on everyday situations to specify what falls inside them and what falls outside.[24]

Driving provides another example of how action in the everyday world, even when it is governed by rules and laws, is not well defined. Although there are explicit rules of the road, there are also numerous unspoken conventions and behaviors, which vary widely from culture to culture: think of driving in Athens or Tel Aviv versus driving in Los Angeles or Manhattan. The challenge of designing self-driving cars that can function in a context-sensitive way indicates the complexities of these unspoken conventions and the impossibility of capturing them in well-defined rules. There is also the problem of how to handle the long list of unlikely, and hence very difficult to foresee, situations that will not appear in the training data for supervised machine-learning systems. Mitchell writes that these are "situations that the vehicle was not trained on, and that might individually occur rarely, but that, taken together, will occur frequently when autonomous vehicles are widespread. . . . Human drivers deal with these events by using common sense—particularly the ability to understand and make predictions about novel situations by analogy to situations the driver already understands."[25] In other words, human drivers realize what is relevant; self-driving cars do not.

One way we to try to get around this problem is by reshaping our environment to make its structure as definite and well defined as possible, so that technologies like self-driving cars can gear into it accordingly. We may think we're making artificial systems with their own genuine autonomy, but we're actually remaking our environment to fit the limitations of our nonautonomous devices, as in allowing self-driving cars to operate only in geo-fenced areas (geographical areas defined by a virtual boundary) with extensive supporting infrastructure.[26] Our inability to see that we have actually been refashioning our life-world to support our devices, rather than creating genuinely cognitive systems, is a case of the Blind Spot.

Assuming that the everyday world consists of definite states that need to be computationally represented inside the head is another instance of the Blind Spot. As the frame problem and the cognitive limitations of deep-learning neural networks indicate, the assumption that the everyday world consists of states with determinate boundaries is a mistake. That assumption is actually a case of surreptitious substitution, of substituting computational states for the imprecise and fluid everyday world.

The Computational Blind Spot

Two interrelated aspects of what we will call the computational blind spot have now come to light. One is surreptitiously substituting computational models for the everyday world; the other is failing to see how we are led to remake the world so that it gears into the limitations of our computational systems. Both prevent us from seeing how our experience of the world relates to our computational constructions.

The computational blind spot lies deep in the history of cognitive science. It was the basic reason for the failure of first-wave, logic-based AI. The

assumption underlying first-wave AI was that the world consists of determinate states, objects, properties, and events (like the positions, pieces, and
moves in a game) and that perception requires having a computational representation of them inside the system. It was assumed that intelligence could
be built into a system by making it perceive "discrete, well-defined, mesoscale
objects exemplifying properties and standing in unambiguous relations."[27]
But we do not perceive the world in these terms. We perceive the world as
made up of open-ended, fluid situations that solicit and afford actions.[28]
If we recall the map and territory metaphor, our sensory perception of the
world is the territory, and computational representations of objects and
events by mathematical functions and binary logic are the map. First-wave
AI tried to build intelligence into a system by giving it a technoscientific
perception of the world, by making it perceive the world in terms of a logic-
based, computational ontology, by making it perceive the world as if the
surreptitious substitution were true. Instead of asking how we come to find
the everyday world perceptually meaningful, first-wave AI started from theoretical representations and abstract reconstructions of the everyday world,
and assumed that the world exists ready-made in those terms.[29] This is a case
of the fallacy of misplaced concreteness, of treating abstract computational
models—the maps—as if they were concretely real—the territory. In short,
first-wave AI used a computational version of the Blind Spot as its blueprint
for building intelligent systems (without knowing it was doing this). First-
wave AI's failures to provide a plausible explanatory framework for the mind
and to build anything remotely resembling genuine intelligence testify to
the deleterious effects of the Blind Spot on the science of the mind.

Second-wave, machine-learning AI escapes this form of the computational blind spot, only to succumb to it in another form. Intelligence is
now seen as the ability to find patterns or statistical correlations in large
amounts of input data, but the input data and learning methods themselves partake of the Blind Spot.

Consider first the part of the machine-learning tradition that recognizes
the computational blind spot in its logic-based, AI form. Biological and
artificial neural network research, going back to the pre-AI cybernetic era,
has long considered intelligence to require being able to realize relevance
in the rich and intricate web of relations that knits together experience
and the world before they get conceptualized in thought and language.
For example, adaptive resonance theory, created by Stephen Grossberg

and Gail Carpenter in the 1980s and continually developed over the past decades, uses unsupervised learning, in which the system has to generate its own categories out of input data with no prior examples, to model the brain's ability to generate perceptual meaning in a dynamic, nonstationary environment (one whose statistical properties change over time).[30] This approach opens a window onto an alternative view of what kinds of things exist relevant to cognition—an alternative cognitive ontology—from the logic-based, computational ontology of first-wave AI.[31] The world does not consist of clear-cut states, objects, properties, and events. We arrive at these kinds of divisions through intellectual abstraction; they populate our specialized models in the scientific workshop, not the world at large. In Smith's words, they are "an *achievement* of intelligence, not a premise on top of which intelligence runs."[32] Substituting one of intelligence's high-level achievements for its underlying constitution is a case of surreptitious substitution. The alternative cognitive ontology is that the world consists of richly differentiated and interwoven elements and relations, which we parse according to our perceptual sensitivities and practical and theoretical projects. Intelligence and relevance realization emerge out of an unconscious background of preconceptual sensemaking supported by our culturally configured and socially situated bodies. It's precisely this perspective that bridges from neural network theory to enactive cognitive science (as we discuss in the next section).

Second-wave AI, however, largely lost sight of this perspective. For example, current deep-learning networks rely on supervised learning methods, in which humans label the training data in advance (tag the data with category names) and supply the system with model input-output pairs so it can learn how to map from new inputs to the right outputs, or on self-supervised learning, in which the network is trained to predict a hidden part of the input sequence. The system not only relies on already codified human knowledge but also winds up encoding historically sedimented human biases. For example, the biases about race and gender in machine-learning training data, particularly in language and image data sets, and the resulting racist and sexist classifications and predictions, have been well documented.[33] Cognitive scientist Abeba Birhane writes, "When ML [machine-learning] systems 'pick up' patterns and clusters, this often amounts to identifying historically and socially held norms, conventions, and stereotypes."[34]

So-called face-recognition systems are a case in point. These systems tend to be more accurate on images of white male faces than on images of female or nonwhite faces. They have classified Asian faces as "blinking" and missed faces with dark skin or, in one notorious case, tagged a selfie of two African Americans as "Gorillas."[35]

Despite their name, face-recognition systems do not, properly speaking, "recognize" faces. As Smith writes, "ML systems learn mappings between (i) images of faces and (ii) names or other information associated with the people that the faces are faces of. We humans often know the referents of the names, recognize that the picture is a picture of the person they name, and so forth, and so the systems can be used *by us* to 'recognize' who the pictures are pictures of."[36] Similarly, large language models (LLMs), which consist of complex statistical models of how the words and sentences in their training data correlate, have no conceptual understanding of the outside world based on real-world experience.[37] Just as AlphaGo has no knowledge that its data structures pertain to a game in the world, and ChatGPT (an AI chatbot built on LLMs) has no knowledge of how the world works, face-recognition systems have no knowledge that their data structures pertain to images of faces, let alone real people who exist in the world.

Besides sorting input into our biased and loaded categories, such as exclusive, discrete, and binary gender or sexual orientation categories, machine-learning systems will also find and learn statistical associations that are irrelevant to the task at hand if they exist in the training data.[38] In the human context, this is a failure of relevance realization.

These limitations of second-wave AI indicate a crucial general point. There is no such thing as a neutral and "value-free" data set.[39] As AI scholar Kate Crawford writes, "Every dataset used to train machine learning systems, whether in the context of supervised or unsupervised machine learning, whether seen to be technically biased or not, contains a worldview. To create a training set is to take an almost infinitely complex and varied world and fix it into taxonomies composed of discrete classifications of individual data points, a process that requires inherently political, cultural, and social choices."[40]

We've been arguing that the blind spot of the AI view of the mind is the experience and understanding of meaning. Here, the surreptitious substitution takes the form of substituting computation for genuine embodied intelligence, the fallacy of misplaced concreteness takes the form of treating abstract computational models as if they were concretely real, the bifurcation of nature takes the form of thinking that human intuition (or the experience of meaning or relevance realization) is a subjective epiphenomenon of objective brain computations, and the amnesia of experience takes the form of forgetting that human experience and biases lie deeply sedimented in AI models.

But the computational blind spot extends further. AI or machine learning has the image, in both popular culture and science, of being a science of "pure intelligence" and an entirely technical domain transcending nature. This image belies AI's fundamental dependence on physical nature and socially organized, collective human knowledge. As Crawford writes, "Artificial intelligence is both embodied and material, made from natural resources, fuel, human labor, infrastructures, logistics, histories, and classifications."[41] AI is fundamentally an "extractive industry," one that "depends on exploiting energy and mineral resources from the planet, cheap labor, and data at scale."[42] On the hardware side, AI requires mining rare earth minerals, such as lithium for batteries, and hence demands oil and coal to power huge mines; on the software side, AI demands huge amounts of carbon-producing energy consumption for running high-performance computer-vision, image-recognition, and language-processing programs. In addition, as we argued earlier and as Crawford notes, "AI systems are not autonomous, rational, or able to discern anything without extensive, computationally intensive training with large datasets or predefined rules and rewards."[43] For these reasons, AI or machine learning in no way transcends nature and in no way constitutes a science and technology of extra-human "pure intelligence." The inability to see these facts clearly and the rhetoric belying them are large-scale effects of the computational blind spot.

Beyond the Computational Blind Spot

We can now come back to the ambiguous relationship between cognitive science and the Blind Spot, introduced at the beginning of this chapter. Cognitive science stands at a crossroads. It can pursue a path that reveals and helps us move beyond the Blind Spot, or it can follow a path that reinforces the Blind Spot. We consider each option in turn.

On the one hand, neural network research, especially when informed by enactive cognitive science, suggests that to be a cognitive system is to make sense out of an immensely intricate, statistically nonstationary, and manifold world rather than to represent (internally mirror) a clear-cut, stationary, and pregiven world.[44] For enactive cognitive science, the living body is the critical node of cognition, and the body's key attribute is being self-individuating. The living body, through the mutual enabling of its parts, makes itself distinct from its immediate surroundings and thereby fashions its environment into a world of relevance. Cognition is sensemaking, to use the enactive slogan. Cognitive scientists Ezequiel Di Paolo, Elena Clare Cuffari, and Hanne De Jaegher write, "Cognition is not about transposing a world of predefined significance into the inside of an agent. It is about agents moving within the world and singly or collectively changing it in ways that are significant according to the forms of life they enact."[45] From this perspective, the significance of research on unsupervised neural networks is that they illuminate how cognition as sensemaking works in mobile and adaptive autonomous agents in a nonstationary and unexpected world (in contrast to a stationary and controlled world). In Grossberg's terms, they illuminate "autonomous adaptive intelligence."[46]

This perspective can help to uncover the Blind Spot by calling attention to how our concepts, classifications, and categories are not ready-made structures mirroring or found in a pregiven world apart from our experience. Instead, they are enacted; they are brought forth by our modes of perception and action and by our social practices. Tracing how this happens via the brain and the rest of the body is a central task of enactive cognitive science. Thus, enactive cognitive science offers ways to help move us beyond the Blind Spot, particularly in its computational form.

On the other hand, when we turn away from this perspective, as typically happens in AI and machine learning, we fall back into the computational blind spot, confusing, as remarked before, the map with the territory. For example, we slide into thinking that the data sets used to train machine-learning systems reflect the world as it really is all by itself at a preconceptual level, and that the system is learning how to categorize and recognize features of that world. We forget that the data sets and training of the system already incorporate human modes of conceptualization, along with their biases, and depend on a massive human infrastructure of knowledge production. We lose sight of the fact that, in Crawford's words, "computational reason and embodied work are deeply interlinked: AI systems both reflect and produce social relations and understandings of the world."[47]

Finally, we forget that AI systems do not know anything about the world as such; instead, they detect statistical correlations in the input we give them with no understanding of what lies behind those correlations.

The stakes are high. Brian Cantwell Smith points out the dangers of what we are calling the computational blind spot: "(i) that we will rely on reckoning systems in situations that require genuine judgement; and (ii) that, by being unduly impressed by reckoning prowess, we will shift our expectations on human mental activity in a reckoning direction."[48] Like Smith, we believe instead "(i) that we learn how to use AI systems to shoulder the reckoning tasks at which they excel, and not for other tasks beyond their capacity; and (ii) that we strengthen, rather than weaken, our commitment to judgment, dispassion, ethics, and the world."[49] This requires recognizing and freeing ourselves from the computational blind spot.

Cognitive science is both uniquely susceptible and uniquely attuned to the Blind Spot. Which path it follows is up to us.

8 Consciousness

Begin Where You Are

Let's begin with an exercise you can perform right now.[1] Use your finger to point to something you see near you. Notice that what you're pointing to has shape, color, and texture. Now point at your foot, and slowly trace your pointing finger up your body, noticing the shapes, colors, and textures of your clothing, limbs, and torso. Finally, turn your finger around, so that it's pointing to where others see your face and directly between your two eyes. Set aside your preconceptions and beliefs about how things are. Instead, just observe carefully how things seem visually. What is your finger pointing at?

Your first inclination probably will be to say your face. But going strictly by how things seem to you visually, does it seem that you're pointing at a face? You see the tip of your finger in the center of your field of vision, but you do not see your face. You don't see any shapes, colors, or textures at the place to which your finger is pointing. You don't see anything there at all. There seems to be an absence of anything visual at that spot.

But that's not quite right either. The visual absence your finger is pointing to seems different from the one over the horizon. That absence lies completely outside your visual field, but your finger seems to be pointing to where your visual field originates. What's being pointed to, though you don't see it, seems present, unlike what lies beyond the horizon, which seems completely absent.

Describing that feeling of presence turns out to be tricky. You might be tempted to say that what feels present is simply you, the one who is looking at the finger. Nevertheless, you don't see yourself at that spot. You don't appear there with any identifying characteristics, unlike when you point to your limbs or torso or to your image in a mirror. Although you visually identify the pointing finger as yours, there's nothing for you to visually identify at the place to which your finger is pointing. The you that feels present there isn't a seen you; rather, it's you as the seer or, more simply, you as seeing. Even the word *you* may say too much, because the feeling of presence seems to be simply a feeling of being aware via seeing. Attaching a personal pronoun to it takes an extra step of thought, one involving the concept of a personal self. Better simply to say that the finger points to the apparent origin or source point of seeing or visual awareness as experienced from within.

You may say, "That source point is just my eye (or a point between them)." But, strictly speaking, that isn't how things seem visually. As Ludwig Wittgenstein writes, "Really you do *not* see the eye. And nothing in the visual field allows you to infer that it is seen by an eye."[2]

Visually speaking, you appear to yourself as headless: there's a gap or opening in place of a head. And yet, this gap isn't a mere vacancy or void. Rather, awareness is there, so it's an "aware-space."[3] That space isn't empty; it's filled with your visual field and everything in it.

Perhaps you'll say, "I may visually appear headless to myself, but I know I have a head, because I can touch it with my other hand [the one whose finger isn't pointing]. And I know my head contains a brain, which is really the origin or source point of my awareness."

But now you've switched from vision to touch. So, if you're going to stay with the spirit of the exercise, you have to observe how things seem just via touch. Attending strictly to how things seem tactilely, is your head any more at the origin of your awareness than your arm or hand, which you can also touch? On the contrary, when you touch your head, the origin or source point of awareness shifts to your hand. Of course, one hand can touch the other one, but the two hands are never simultaneously touching and touched in relation to each other; instead; they alternate their roles spontaneously.[4] Just as the eye does not see itself in its visual field, the hand does not touch itself in its tactile field. As for your brain, it doesn't show up anywhere directly in your experience, not even as an absence with a peculiar felt presence, like the apparent origin of the visual field. If nothing in the visual field allows you to infer that it's seen by an eye, certainly nothing in the visual field allows you to infer it's seen by a brain.

The finger-pointing exercise, which comes from English writer Douglas Harding, is meant to induce a reversal of attention.[5] Ordinarily the arrow of attention points out toward the world, which includes your body. Our attention is typically invested in the world, in the various objects of awareness, and jumps around from one thing to the next. It takes a special effort to turn attention around toward awareness. The finger-pointing exercise is a trick for doing this. It brings awareness into relief and makes it salient. The exercise is an aid to noticing that there's an experiential difference between paying attention to one or another object of awareness, even to your body, which you also experience subjectively, and attending to awareness. Consciousness, understood as awareness and its objects, is our topic in this chapter.

Transparency

Because the topic of consciousness is mired in controversy, we already need to deal with an objection to our way of delineating consciousness. Some philosophers have denied that awareness can be experientially distinguished from the objects of awareness.[6] They say that awareness is transparent, by which they mean that you see right through it to its objects, like seeing through a clear window pane to what lies outside.[7] For example, when you see a sunset, they claim that you're aware only of the colors of the sky, which are perceptible qualities of the object, but not of any qualities of your seeing or visual awareness. In addition, if you try to pay attention to your visual awareness, all you can find are qualities of what you see, but not any qualities of your seeing. Or when you listen to a melody, you can attend only to the notes, but not to any qualities of your listening. In general, according to this viewpoint, awareness and attention make no contribution to how experience feels or seems apart from the objects and properties they present.

We think this viewpoint is wrong.[8] When you see your finger pointing to where others see your face, you're aware of the finger, but you're also aware of seeing. Of course, if you're paying attention to the finger, then your awareness of seeing isn't explicit. It's implicit or tacit. But you can mentally direct your attention to visual awareness, as the finger-pointing exercise illustrates. When you do so, the feeling of being aware intensifies, regardless of the object.

A similar intensification of awareness occurs in a lucid dream. When you realize you're dreaming, you become aware of your dream state as such, and you can direct your attention precisely to the dream awareness, regardless of whatever you happen to be dreaming about.[9] The finger-pointing exercise induces an analogous lucidity in perception, so that you can direct your attention to perceptual awareness as such, regardless of whatever you happen to be perceiving. From this "witnessing" perspective, there is a clear difference between experiential qualities of awareness, such as stability, tranquility, and alertness, versus experienced qualities of the object, such as color.[10]

Asian meditative traditions are relevant here. In certain styles of Buddhist mindfulness meditation practice, you initially learn to anchor your attention to an experiential object, such as the breath, while being vigilant for distraction.[11] Vigilance requires being aware of features of the ongoing experience that pertain not to the object but to the qualities of awareness, such as its stability, degree of alertness, and affective tone (whether it feels pleasant, unpleasant, or neutral). In psychological terms, such vigilance requires a kind of meta-awareness—an awareness of awareness. Crucially, for these styles of practice, cultivating meta-awareness does not involve making an explicit, inward, or introspective turn, because that would mean making awareness the object. Instead, after you've achieved a degree of attentional stability, you drop attention to the anchor (the experiential object) while remaining in a state of "mere nondistraction." This attitude heightens the salience of off-object qualities of awareness (qualities of awareness that aren't tied to the object) so they can be noticed without focusing on awareness as an object. You're able to spy on awareness out of the corner of your eye, as it were, without explicitly targeting it. You can sustain a nonobjectifying kind of meta-awareness, "the spy of mindfulness," that doesn't present awareness as an object.

This perspective is easily related to the finger-pointing exercise, especially since Harding himself was writing from a similar kind of meditative perspective.[12] The finger serves as the initial attentional anchor. But because of the peculiar way it's pointing, it effects a reversal of attention away from itself to awareness. What thwarts turning awareness into an object is the experience of being "headless," the encounter with the "aware-space" that isn't an object and that's filled with the visual world. Awareness doesn't show up apart from the objects of awareness, yet it's clearly distinguishable from them. Eventually, after sensitizing yourself to awareness in contrast to its objects, you don't need to rely anymore on the finger-pointing prop. Instead, you can "just sit," to borrow Zen philosopher Dōgen's words, in a state of open awareness.[13]

The Epoché

Let's now take the finger-pointing exercise in reverse. Starting from the origin of your visual field, you look down and see your torso, arms, and legs, extending out from the apparent source point of awareness via vision. Now your body, and indeed everything else you see, seems contained within awareness. As Robert Sharf writes, "The topography of this space is akin to the twisting of a Möbius strip, which is simultaneously one-sided and two-sided, which is to say that it is impossible to determine where the subject ends and the object begins."[14] At the same time, awareness is intrinsically open to what lies beyond it, to the world outside its range, so awareness exhibits both closure (like a Möbius strip or Klein bottle) and transcendence (it intrinsically points beyond itself and constantly expands its range).[15]

When we look at things this way, we are in effect carrying out a version of what phenomenologists, following Husserl, call the *epoché*, a term that comes from ancient Greek skepticism and means a suspension of judgment or withholding of assent.[16] Ordinarily we judge that objects exist "out there" independent of us. Phenomenologists call this perspective the "natural attitude." The first step to adopting a properly phenomenological attitude toward experience or consciousness is to refrain from the natural attitude. The point isn't to deny the natural attitude, but instead to adopt a different way of looking at things. The phenomenologist brackets the everyday positing of the world as existing outside consciousness in order to examine the world strictly as it is disclosed to consciousness. The aim is to investigate things exactly as they are correlated with how we consciously apprehend them. The epoché is the mental act of suspending our reliance on the natural attitude, which takes objects for granted as "just there," so that attention can be directed to exactly how things show up within conscious awareness. This kind of attention focuses on the correlational structure of consciousness and its objects (awareness and objects of awareness). Thus, the epoché requires essentially the same kind of meta-awareness (awareness of awareness) that figures in the finger-pointing exercise and certain kinds of meditation.

In the natural attitude, we take reality for granted as being simply there. Looked at phenomenologically, however, reality is that which is disclosed to us as real, whether in everyday perception or scientific investigation. The idea is not that reality would not exist if not for consciousness or that reality is made out of consciousness. The idea is that we have no grip on what reality means apart from our ability to apprehend something as real, and such apprehension essentially involves the workings of consciousness. The epoché is a philosophical and "spiritual exercise" for gaining access to consciousness as a condition of possibility for the intelligibility of reality, for the very idea of reality to be meaningful.[17]

The Primacy of Consciousness

The epoché brings to the fore what we will call the primacy of consciousness.[18] There is no way to step outside consciousness and measure it against something else. Everything we investigate, including consciousness and its relation to the brain, resides within the horizon of consciousness.

The horizon metaphor is significant. The line of the horizon is a limit beyond which we cannot go, because it keeps receding as we approach it. It's an apparent line and travels with us. In one sense, the horizon is "out there": it's the farthest point the eye can see before the Earth's surface curves away beneath our view. In another sense, the horizon is "in here": it's a structure of perception and doesn't exist apart from the mind.

The horizon metaphor reveals a way of thinking about consciousness that we will call the horizonal conception.[19] Consciousness is the horizon for the world's disclosure: The world appears and is present to us from within the horizon of consciousness. Consciousness is the horizon of anything we can perceive, think about, and investigate, including when we are doing science. We can observe, imagine, and investigate things only within the horizon of consciousness, and anything that we determine to be real or factual gets this determination from within the horizon of consciousness.

Most scientists and philosophers do not look at consciousness this way. Instead, consciousness is regarded as just another phenomenon in the world. Of course, theorists disagree about the nature of this phenomenon. Some hold that consciousness is produced by the brain or is identical with a state of the brain; others hold that it's something nonphysical, irreducible to physical nature. Some say that it's a kind of cognitive illusion, while others argue that it's one of the fundamental ingredients of reality, which a purely physical description cannot capture. Nevertheless, despite these differences, the common idea is that consciousness is a phenomenon in the world and that it can be distinguished from other phenomena.

The horizonal conception is different. Instead of being just a phenomenon in the world, consciousness is also understood as a precondition of the world's disclosure. Consciousness is the horizon within which any phenomenon we can talk about or point to is present. This includes anything we may be able to indicate about particular states of consciousness in relation to the brain. Consciousness as a phenomenon in the world always shows up inside consciousness as a horizon.

At the same time, horizonal consciousness is nothing in itself or all by itself, because it's nothing other than the disclosure or manifestation of the world. Disclosure and manifestation imply a recipient, the one to whom something is disclosed or manifested. For this reason, phenomenologists often describe consciousness as the "dative of manifestation" (manifesting is manifesting *to*). Consciousness is the recipient of the manifestation of presence, including all the ways that absences, such as what's over the horizon, are intimated within presence. Using Buddhist philosophical terms, we can say that horizonal consciousness is "empty" of "own-being," because it's "dependently arisen" with the manifestation of the world.

We can now say more fully what we mean by the primacy of consciousness.[20] Horizonal consciousness is not something we have, in the way we have a feeling of pain or a visual impression of a color. It's something we live. It's a mode of being or a form of existence. So it has *existential primacy*. Consciousness also has *cognitive primacy*. As our parable of temperature illustrates and as we have argued throughout this book, conscious experience is both the point of departure and the ultimate source of validation for acquiring knowledge, particularly scientific knowledge. Finally, consciousness has *transcendental primacy*. Something is transcendental if it is an a priori condition of the possibility of knowledge (a condition not provided by experience but presupposed by it). Consciousness is not just another object of knowledge, but also, and more fundamental, that by which any object is knowable. For this reason, consciousness is irreducible to any object or domain of objects: any explanation of consciousness in terms of a specific object, such as the brain, or even a totality of objects, already presupposes consciousness as that by which objects are individuated and intelligible.

As we said at the outset of this section, there is no way to step outside of consciousness and account for it exhaustively in terms of something else.

Some thinkers, ancient and modern, have gone on to infer from these kinds of thoughts that the universe, nature, or reality *is* essentially consciousness or is somehow made out of consciousness. But this does not logically follow. This further thought takes a speculative leap beyond what we can know or establish on the basis of the foregoing considerations about consciousness as experienced from within and as an irreducible precondition of scientific knowledge. Furthermore, as we will now see, this speculative leap runs afoul of what we will call the primacy of embodiment, which is as equally undeniable as the primacy of consciousness.

The Primacy of Embodiment

Earlier we said that your brain doesn't show up anywhere directly in your experience. We can now turn this point around. There is an undeniable, close dependency of states of consciousness on the brain, but this dependency is inscrutable from within the horizon of first-person awareness. Being awake and perceiving things around you, being asleep and dreaming, "just sitting" in open awareness and mere nondistraction: these modes of awareness depend on the brain, but experiencing them from within reveals nothing of this dependence. More generally, horizonal consciousness does not directly reveal from within all that it depends on.

Of course, some philosophical and spiritual traditions deny this. They assert that it's evident that the inner nature of consciousness, sometimes called "pure consciousness" or "pure awareness," depends on nothing other than itself, and hence constitutes the fundamental ground of being. But if this metaphysics is thought to follow from the experience of pure awareness, the inference is invalid. No matter how basic and self-subsistent awareness may seem to be from the inside, it doesn't follow that it really is this way. How something appears doesn't necessarily reveal how it is, even in the case of consciousness. This doesn't mean that there aren't genuine experiences of "pure awareness," states in which awareness is experienced as having an invariant phenomenal quality distinct from its variable objects. But it does mean that such experiences, by themselves or taken just on their own terms, don't reveal their origins or sources, or all that they depend on.

We can take this point further. Consciousness, in the sense of your immediate awareness, must have sources outside itself. Anything you can think, including "this finger is pointing to an aware-space" or "this is a dream," is meaningful thanks to concepts you possess and know how to use, and whose meanings don't originate with you. Anything you do, including even "just sitting" and "nonthinking" in meditation, has meaning, and is possible at all, only given a background of social and cultural practices that extend far beyond you, and that predate and will outlive you. Even meta-awareness, the ability to mentally attend to and monitor your awareness, originates developmentally from your having internalized an outside perspective on yourself when you were an infant interacting with others. Finally, any attempt to deny that your immediate awareness has outside sources is incoherent, because this denial cannot be thought or stated except through the conceptual and linguistic resources provided to you by your social heritage and your cultural and biological evolution. Even if you were the last surviving person of a universal plague, your mind and your capacity for meta-awareness would be inherently social.

These points describe what we will call the primacy of embodiment. There is no way to step outside embodiment, our living and inherently social and encultured bodies. As Merleau-Ponty writes, "The body is the vehicle of being in the world."[21] The body is the medium of the world's disclosure via consciousness.

A Strange Loop

We're now confronted with a strange loop.[22] Horizonal consciousness subsumes the world, including our body experienced from within, while embodiment subsumes consciousness, including awareness in its immediate intimacy. The primacy of consciousness and the primacy of embodiment enfold each other. We need to examine this strange loop, which disappears from view in the Blind Spot.

In fact, we've already encountered this strange loop in our discussion of time and cosmology and our discussion of life: Our experience of time depends on the flow of cosmic time that we measure through our experience of time, and only life can know life. Like the ouroboros, the serpent swallowing its own tail, we are in the universe and the universe is in us. This is the strange loop.

Merleau-Ponty puts his finger on the strange loop when he writes in *Phenomenology of Perception*: "The world is inseparable from the subject, but from a subject who is nothing but a project of the world; and the subject is inseparable from the world, but from a world that it itself projects."[23] This statement is meant to clear a path between two extremes. One is the idea that there is a world only for or in consciousness (idealism). The other is the idea that the world exists ready-made and comes presorted into kinds or categories apart from experience (realism). Instead of these two extremes, Merleau-Ponty proposes that each one of the two terms, the conscious subject and the world, makes the other one what it is, and thus they inseparably form a larger whole. In philosophical terms, their relationship is dialectical.

The world Merleau-Ponty is talking about is the life-world, the world we're able to perceive, investigate, and act in. The subject projects the world because it brings forth the world as a space of meaning and relevance. But the subject can project the world only because the subject inheres in a body already oriented to and engaged with a world that surpasses it. The bodily subject is not just in the world but also of the world. The bodily subject is a project of the world, a way the world locally self-organizes and self-individuates to constitute a living being.[24]

You may want to say that the universe—the whole cosmos or all of nature—subsumes the life-world, so the strange loop pertains only to us and our life-world, not to us and the universe altogether. But quarantining the strange loop this way won't work. It's true that our life-world is a minuscule part of an immensely vaster cosmos. The cosmos contains our life-world. But it's also true that the life-world contains the universe. What we mean is that the universe is always disclosed to us from within the life-world. The life-world sets the horizon within which anything is observable, measurable, and thinkable. So the life-world and the universe themselves are caught up in a strange loop.

Merleau-Ponty has this strange loop in mind when he writes:

> For what exactly is meant by saying that the world existed prior to human consciousnesses? It might be meant that the earth emerged from a primitive nebula where the conditions for life had not been brought together. But each one of these words, just like each equation in physics, presupposes *our* pre-scientific experience of the world, and this reference to the *lived* world contributes to constituting the valid signification of the statement. Nothing will ever lead me to understand what a nebula, which could not be seen by anyone, might be.

> Laplace's nebula is not behind us, at our origin, but rather out in front of us in the cultural world.[25]

Merleau-Ponty is not denying that there is a perfectly legitimate sense in which we can say that the world existed before human consciousness. Indeed, he refers to the "valid signification" of this statement. He is making a point at a different level, the level of meaning. The meanings of terms in scientific statements, including mathematical equations, depend on the life-world, as our parable of temperature and our discussion of the dependence of clock time on lived time illustrate. Furthermore, the universe does not come ready-made and presorted into kinds of entities, such as nebulae, independent of investigating scientists who find it useful to conceptualize and categorize things that way given their perceptual capacities, observational tools, and explanatory purposes in the life-world and the scientific workshop. The very idea of a nebula, a distinct body of interstellar clouds, reflects our human and scientific way of perceptually and conceptually sorting astronomical phenomena. This is what Merleau-Ponty means when he says that he cannot understand what a nebula that could not be seen by anyone might be. Nothing intrinsically bears the identity "nebula" within it. That identity depends on a conceptual system that informs (and is informed by) observation. Nevertheless, Merleau-Ponty's last sentence is exaggerated. Given the conceptual system of astrophysics and general relativity theory, Laplace's nebula *is* behind us in cosmic time. But it is not *just* behind us. It is also out in front of us in the cultural world, because the very idea of a nebula is a human category. The universe contains the life-world, but the life-world contains the universe. This is the strange loop.

We can now appreciate that the life-world has the same kind of primacy as the primacy of consciousness and the primacy of embodiment. Better yet, the primacy of the life-world subsumes the primacy of consciousness and the primacy of embodiment. We cannot step outside the life-world, because we carry it with us wherever we go. Whatever we know about the universe is known from within the life-world. At the same time, we constantly open up new vistas with new life-world horizons, particularly in science (but also in art, religion, and philosophy). So there must always be that which transcends the life-world, that which lies beyond it and provides a source of ever new possibilities of disclosure and manifestation from within it. We can call that source "nature."[26] The life-world is inherently open to this source that transcends it. The life-world exhibits both closure (any action within it results in another action within it) and transcendence (it is intrinsically open to what lies beyond it). The life-world and the universe enfold each other.

The Blind Spot hides this strange loop so that we forget it. We come under the spell of thinking that it's possible to grasp consciousness from a purely outside vantage point, one outside the strange loop in which consciousness places us. In other words, we think we can comprehend consciousness within the framework of reductionism, physicalism, and objectivism or, failing that, by postulating a dualism of physical nature versus irreducible consciousness that we could somehow grasp outside the strange loop. We're now ready to see why these ways of thinking won't work and why the reasons for their failure go deeper than what philosophers call the hard problem of consciousness.

The Hard Problem Is an Artifact of the Blind Spot

The hard problem of consciousness is said to be the problem of explaining how a physical system, such as the brain, gives rise to conscious experience.[27] The expression "gives rise to" is vague, because part of the problem is to figure out exactly what the explanatory relation between a physical system and conscious experience is supposed to be. Although "gives rise to" suggests a causal relation, maybe the relation is identity (being conscious is identical to a certain kind of physical state) or realization (being conscious is realized in a certain kind of physical state). In any case, the problem, in general terms, is to explain how and why conscious experience is generated in a physical system, as described in the objective terms of science, whether that description be in terms of electrochemical processes in neurons and synapses, computational or informational processes in neural networks, or quantum effects in biological systems or matter in general.

Today's hard problem of consciousness is an artifact of the Blind Spot. The problem is part and parcel of the mind-body problem that arose with modern science. Already in 1866, Thomas Henry Huxley, the nineteenth-century English biologist known as "Darwin's Bulldog," wrote, "But what consciousness is, we know not; and how it is that anything so remarkable as a state of consciousness comes about as the result of irritating nervous tissue, is just as unaccountable as the appearance of the Djin when Aladdin rubbed his lamp in the story, or as any other ultimate fact of nature."[28] A few decades later, William James, in his *The Principles of Psychology* (1890), quoted words to the same effect from Irish physicist John Tyndall's 1868 presidential address to the Physical Section of the British Association for the Advancement of Science: "The passage from the physics of the brain to the corresponding facts of consciousness is unthinkable. Granted that a definite thought, and a definite molecular action in the brain, occur simultaneously; we do not possess the intellectual organ, nor apparently any rudiment of the organ, which would enable us to pass, by a process of reasoning, from the one to the other."[29] Why should brain processes eventuate in any conscious experience at all? This is the hard problem of consciousness, also known as the explanatory gap.[30]

The problem goes back to the rise of modern science in the seventeenth century, particularly to the bifurcation of nature, the division of nature into external, physical reality, conceived as mathematizable structure and dynamics, and subjective appearances, conceived as phenomenal qualities lodged inside the mind. The early modern version of the bifurcation was the division between "primary qualities" (size, shape, solidity, motion, and number), which were thought to belong to material entities in themselves, and "secondary qualities" (color, taste, smell, sound, and hot and cold), which were thought to exist only in the mind and to be caused by the primary qualities impinging on the sense organs and giving rise to mental impressions. This division immediately created an explanatory gap between the two kinds of properties. Thus, John Locke, in his monumental 1689 book *An Essay Concerning Human Understanding*, posed the problem that Huxley would restate roughly two hundred years later. In Locke's words, "We are so far from knowing what figure, size, or motion of parts produce a yellow colour, a sweet taste, or a sharp sound, that we can by no means conceive how any *size, figure, or motion* of any particles, can possibly produce in us the idea of any *colour, taste, or sound* whatsoever; there is no conceivable connexion betwixt the one and the other."[31]

Gottfried Wilhelm Leibniz (whose many achievements include independently creating calculus around the same time as Newton, inventing mechanical calculators, documenting and refining the binary number system, and being probably the first European philosopher to write seriously and sympathetically about Chinese philosophy) raised a similar problem in his *Monadology*, published in 1714: "If we imagine that there is a machine whose structure makes it think, sense, and have perceptions, we could conceive it enlarged, keeping the same proportions, so that we could enter into it, as one enters into a mill. Assuming that, when inspecting its interior, we will only find parts that push one another, and we will never find anything to explain a perception."[32] "Perception" was Leibniz's general term for mental states. He argued that when we inspect a mechanical system, which is easier to do when we imagine it enlarged so that we can examine its interior, we will find nothing that explains the presence of mental states (sensations, perceptions, and thoughts).

Although Leibniz's mill argument was not specifically about consciousness but instead about mental states in general, it's easy to imagine a contemporary version of his argument applied to the brain and consciousness. We're now able to enter the brain and see its interior structure and dynamics thanks to cellular and molecular neurophysiology, and neuroimaging technologies. We now have many "connectomes," comprehensive maps or wiring diagrams of neural connections in the brain. Is there anything in such models of the brain that explains the presence of consciousness? Suppose we were able to image the physical activity of the neural networks that make up your brain at whatever spatial and temporal scale we choose. Although we'd be able to establish precise correlations between your brain activity and your experiences, these correlations aren't enough to explain your experiences, including both the fact of their occurrence and their particular qualities. The physical and computational activity of the neural networks lies on the other side of a conceptual and explanatory divide from your direct experience of an azure sky, a summertime reverie, or quiet sitting in meditation. This chasm is the explanatory gap.

Despite the amazing, nonstop advances in physics, biology, and neuroscience, no fundamental progress on bridging the chasm between consciousness and physical models has been made in science since the bifurcation of nature that began with the rise of modern science. Although physical and biological models are increasingly sophisticated and informed by increasing amounts of data, the chasm remains. The problem that Huxley and Tyndall highlighted in the nineteenth century is the same one that philosophers Thomas Nagel and David Chalmers identified in the twentieth century and persists today.[33] Indeed, it is hard to see how any advance in understanding physical processes, described in completely objective terms at whatever scale or level, will allow us to bridge this chasm. This situation should lead us to suspect that the hard problem of consciousness is built into blind-spot metaphysics, and not solvable in its terms.

As we've seen throughout this book, blind-spot metaphysics arises when we mistake a method for the intrinsic structure of reality. We devise a powerful explanatory method that abstracts away from consciousness while forgetting that the method remains fundamentally dependent on consciousness. As our parable of temperature and discussion of time illustrate, we begin by replacing concrete and tangible phenomena with abstract and idealized mathematical constructs (the surreptitious substitution). Because these constructs are so useful, we forget they are abstractions and take them instead to be concrete reality (the fallacy of misplaced concreteness). As we ascend the spiral of abstractions, we come to regard what are really highly distilled residues of experience as if they were essentially nonexperiential entities that constitute the objective fabric of reality (the reification of structural invariants). Finally, conscious experience, the method's source and touchstone, drops out of sight completely, hidden in the Blind Spot (the amnesia of experience). The result is a picture of reality drawn in physicalist, reductionist, and objectivist terms, from which consciousness has been excluded by construction.

If we now turn around and try to recuperate consciousness within the blind-spot worldview, using the same method by which we excluded it at the outset, we face an impossible and nonsensical situation. Claiming that consciousness can be reductively explained by its abstract structural residues in physics or cognitive neuroscience inverts the whole procedure of generating scientific knowledge, which starts from and is forever beholden to direct experience. The move is absurd in principle because it tries to replace the subjectivity it excluded at the start with an abstraction cast in completely objective terms.[34] It fails to recognize the ineliminable primacy of consciousness in knowledge.

There are only a few options for trying to deal with consciousness within the confines of the blind-spot worldview. Ultimately, they're all unsatisfactory, because they never come to grips with the need to recognize the primacy of consciousness and the strange loop in which we find ourselves. The first option is to change the subject by focusing on a different problem, such as which aspects of conscious experience go along with which aspects of brain activity. The second option is to entrench the bifurcation of nature in a metaphysical dualism of physical reality and irreducible mental properties. Depending on how widespread these mental properties are thought to be, the resulting position is either naturalistic dualism, which postulates additional bridging principles between physical processes and irreducible mental properties to explain how conscious experience arises from the brain, or panpsychism, which takes consciousness to be a fundamental feature of the physical world.[35] The third option is to say that consciousness is an illusion, a position known as illusionism. Regardless of which approach we choose, however, we stay within the ambit of the Blind Spot, by doing more of the same "normal science," by injecting a dualist or panpsychist "extra ingredient" into an otherwise unchanged and preserved blind-spot worldview, or refusing to countenance consciousness except as a kind of cognitive illusion, a move that exacerbates the Blind Spot.[36] In the rest of this chapter, we take a closer look at these options and explain why we need a radically different approach beyond the Blind Spot.

Banking on the Brain

The standard approach in the neuroscience of consciousness is to ignore the hard problem of how consciousness is generated in the physical world, and to focus instead on mapping experiential properties of consciousness onto functional properties of the brain.[37] The term *neural correlates of consciousness* refers to this kind of mapping.[38] For example, the properties of being awake and consciously perceiving visual detail are known to be closely correlated with large-scale, coordinated patterns of neural activity across widespread cortical areas.[39] Similar large-scale brain networks are active when we dream but not during deep sleep.[40]

The brain topography of these networks, however, remains a matter of debate. Are the brain areas most closely correlated with conscious perception the sensory ones at the back of the head or the cognitive ones at the front of the head?[41] The debate also concerns whether perceptual consciousness is essentially a sensory phenomenon or a cognitive one comprising attention and memory. The issue has crystallized into a debate between "no report" versus "report" paradigms.[42] The traditional way to investigate the neural correlates of consciousness is to use verbal reports about perception as the standard for determining the contents of consciousness. But verbal reports confound the contents of perception with the cognitive processes of attention and memory that are required to make those contents reportable (you have to pay attention to something and hold it in memory to report it). No-report paradigms try to remove these confounds by first using reports to find neurophysiological and behavioral indicators of consciousness (such as electrophysiological and brain-imaging measures, together with eye movements and pupil dilation) that subsequently can be used in the absence of reports. Each side in this debate, however, has been able to provide empirical evidence for its position (for conscious perception either being essentially tied to attention and memory or not dependent on them). As a result, there is no single, generally agreed-on neural correlate of consciousness, and there remains an apparently unresolvable disagreement about what consciousness is (whether it's fundamentally a cognitive capacity or a sensory one).

Another experiential property of consciousness is the sense of self, the feeling of being a subject of experience and an agent of action. The sense of self is not one thing. It includes many different elements, such as the bodily self, the mental self experienced in memory and anticipation, the self as a personal storyline (the narrative self), and the social self. Each of these aspects in turn can be analyzed into distinct facets. For example, the bodily self includes the feelings of ownership ("this body is mine"), agency ("I'm the one making this movement"), self-location ("I'm in my body"), and egocentric perception ("I'm seeing the world through my own eyes"). The sense of self is a construction, made from integrating different kinds of self-experience. Their characteristics and integration can also be mapped onto various systems in the brain and the rest of the body.[43]

Mapping from properties of consciousness to properties of the brain has produced fruitful models and experimental findings. It's an important research strategy in the neuroscience of consciousness. The problem arrives when it's said that this approach will eventually make the hard problem or the explanatory gap go away.[44] This is a mistake. To establish detailed mappings from experience to the brain is one thing. To explain what makes these mappings possible, and how and why they hold, is another thing. As Michel Bitbol remarks, thinking that solving these mapping problems will finally solve (or dissolve) the hard problem is like thinking you can reach the horizon by walking far enough.[45] No matter how detailed the mappings between experience and the brain are, they don't suffice to explain the physical origin or generation of consciousness. Conscious experience is a given of the mapping relation; it's not explained by it. Accounting for the experiential side of the mapping requires a different kind of explanation.

Research on the neural correlates of consciousness faces additional problems when it's taken as a strategy for dealing with the hard problem or the explanatory gap. Unless we know the precise nature and structure of consciousness, we can't determine the conditions under which it occurs. Could consciousness in some form exist in organisms without a brain or nervous system?[46] We have no way of definitively answering this question at the present time. Although we know that the brain is necessary for the cognitive capacities that are closely associated with consciousness in us, we don't know whether these cognitive capacities are essential to consciousness or whether they're extraneous. Research on the neural correlates of consciousness has been unable to decide this matter, as we've seen. Consider sentience, the capacity to feel. Does it require cognitive capacities like selective attention and short-term memory, or is it separate from them? Is there a minimal or core form of sentience, a feeling of being alive, that belongs to all organisms? Or does sentience emerge at some point in evolution? There is no consensus on these questions in the science of consciousness. Even if many neuroscientists would say that consciousness is restricted to animals with brains, this position is at best a working assumption or hypothesis, not an established fact.[47]

If we already knew which experiential and cognitive properties were necessary for consciousness and we knew that they were exclusive to the brain, then mapping them onto the brain would help us zero in on the physical origin of consciousness, even if we still didn't understand how a physical system such as the brain could generate consciousness. But we don't know these things and a challenging circularity arises when we try to figure them out.[48] To determine the extent of consciousness in the natural world, we need a validated theory of consciousness, but to arrive at such a theory, we need to rely on assumptions about how widespread in nature consciousness is. Having a validated theory of consciousness requires determining which experiential and cognitive properties are essential to consciousness, but determining this depends on assumptions about which kinds of organisms are conscious. We can hardly ignore these challenges if we're trying to answer the question of how consciousness arises in nature.

Some theorists make an analogy between the neuroscience of consciousness and the biology of life.[49] Just as life isn't one thing—it includes metabolism, growth, self-maintenance, reproduction, evolution, and so on—so consciousness isn't one thing. And just as progress was made on understanding life by explaining the different properties of life, so progress on consciousness will be made by mapping its different forms and aspects onto the brain.

But this analogy only goes so far. It's not the case that the problem of life has been solved by the sort of mechanistic understanding that advocates of the mapping strategy for the brain and consciousness have in mind. On the contrary, accounting for organisms as self-determining agents with unprestateable phase spaces (see chapter 6) has barely begun. Furthermore, the impression that biology has completely solved the problem of life partly derives from our having excised the problem of consciousness from the problem of life. But this just kicks the can down the road. Being conscious is part and parcel of life-regulation processes in conscious organisms.[50] We need to explain how the emergence of conscious organisms happened, including whether sentience is coeval with life or a later evolutionary event, and whether the emergence of consciousness was accidental or inevitable from an evolutionary standpoint. So the problems of life and consciousness intertwine. Furthermore, once we try to remove the problem of consciousness from the problem of life, the disanalogy between them becomes salient. Consciousness is subjective and experiential, whereas biological life is construed as entirely objective. The disanalogy isn't just that we have to collect data about subjective experiential states in the case of consciousness, but not in the case of life. It's also that the phenomena are conceptualized in different ways. In other words, we use different conceptual frameworks to understand them—the framework of objective structure, function, and dynamics, on the one hand, and the framework of subjective experience, on the other. Yes, we can map between these two frameworks. But this still leaves untouched the questions about what makes the mapping possible and how to account for the generation of consciousness in the natural world.

The Predicting Predicament

"Predictive processing," a currently popular theory of how the brain works as a cognitive system, has been proposed as a way to answer these questions and explain how the brain generates perceptual experience.[51] According to this theory, the brain constantly makes predictions about the sensory signals it receives, based on hypotheses about the hidden, unobservable causes of the signals "out there" in the world. The brain works to minimize the differences between its predictions and incoming sensory signals. Sensory signals serve as prediction errors that the brain uses to update and revise its hypotheses and predictions, which it applies to the next round of signals. This cycle of prediction, hypothesis updating, and new prediction is thought to work according to the principles of Bayesian probability theory (discussed in our treatment of QBism in chapter 4), so this viewpoint is also described as the "Bayesian brain" hypothesis. In short, according to the theory, perceiving results from the brain's making and revising predictions in the face of incoming sensory signals treated as error signals, and what we perceive is the content of the brain's predictions as it works to minimize sensory prediction errors.[52]

Predictive processing was developed as a theory about how the brain functions in perception, not as a theory about consciousness. But the idea that what we perceive, the content of our perceptual experience, is the content of our brain's internal predictive model is meant to provide a link to conscious experience. According to this viewpoint, we don't perceive the world directly; instead, we experience the content of our brain's model of the outside world. Neuroscientist Anil Seth describes this idea by saying that perceptual experience is "controlled hallucination," a phrase that goes back to the twentieth-century British AI researcher Max Clowes, and possibly to the nineteenth-century German physicist and physician Hermann von Helmholtz.[53] (Neuroscientist Rodolfo Llínas, in 1994, also described perception as controlled dreaming—dreaming with ongoing, real-time sensorimotor constraints.[54]) The basic idea is that hallucination happens when the brain's predictions or expectations dominate perception while being unconstrained by sensory signals. Perception happens when the brain's predictions are tightly constrained by sensory signals, so that the brain constantly revises its predictions in light of them. The idea that perception is controlled hallucination (and hallucination is unconstrained perception) has been offered as a way to extend predictive processing from a theory about how brains work to a theory about how the brain's workings account for perceptual experience.

But saying that conscious perception is the brain's controlled hallucination doesn't really explain consciousness. Although the predictive processing theory revises the terms on the brain side of the experience-brain mapping, it doesn't explain how the brain, understood in predictive-processing terms, suffices for the first-person experience of being aware. It's true that the theory requires more than determining which patterns of brain activity go with which conscious experiences. We also have to determine which aspects of the brain's predictive processing go with which conscious experiences. Furthermore, we can gain insight into the experience-brain mapping by manipulating the brain's predictive processing and thereby altering experience. We can also manipulate subjective experience to see how predictive processing changes. Still, why should a predictive-processing system embody an experiential point of view? Why should it be consciously aware? What is it about predictive processing that suffices for consciousness? The theory doesn't answer these questions.

Let's be clear about the issue here. It's not whether the predictive processing theory provides useful models of brain function. Clearly, it does. Even so, claiming that the brain is nothing but a predictive processing system is overblown, given the current state of the evidence and the methodological challenges in testing the models.[55] The "Bayesian brain" concept is also inflated. As neuroscientist Luiz Pessoa remarks, no one calls the brain the "Newtonian brain" because we can use differential equations to model it, so just because we can use Bayesian probability theory to model aspects of brain function doesn't make the brain a "Bayesian brain."[56] The map is not the territory, and neither is the math.[57] Nor is the issue whether the predictive processing theory is useful for identifying and refining our understanding of the neural correlates of consciousness.[58] The issue is whether the theory explains consciousness by explaining how the brain generates experience. Does the theory solve the "generation problem" for consciousness—the problem of how a physical system can produce conscious experience?[59] The answer is no.

We feel compelled to insist on this point, because some neuroscientists claim otherwise. They slide from saying that the predictive processing theory is useful for refining the experience-brain mapping to claiming that it provides a mechanistic model of what conscious experiences are and how they're generated in the natural world. For example, we're told that our conscious perception of the world is "nothing more and nothing less than our brain's best guess of the hidden causes of its colorless, shapeless, and soundless sensory inputs," and that "the specific experience of being you (or me) is nothing more than the brain's best guess of the causes of self-related sensory signals."[60] These "nothing more" statements go far beyond anything the predictive processing theory scientifically establishes. They're not scientific statements; they're metaphysical ones. They fly in the face of the explanatory gap between predictive processing models of the brain and the person's conscious experience. They exemplify the Blind Spot.

Saying that perceptual experience is the brain's best guess about the hidden outside world creates intractable problems, ones that typify the Blind Spot. We're stuck predicting inside our heads. The job of our senses isn't to disclose the world, including our bodies, but rather to indicate errors in what we guess to be the case based on our skull-bound models. The predicting predicament is essentially solipsistic: immediate experience presents us with just our own, private, internal predictions, which we update only by registering our errors about a forever-hidden outside world. This is a cognitive-neuroscience version of the bifurcation of nature.

There is nothing scientifically mandatory about this picture of perception. On the contrary, other scientific theories of perception reject it. The ecological psychology of James J. Gibson and his followers understands perception as an activity of the whole animal, not an episode inside its head.[61] The content of perception is the layout of the environment and what that layout affords the animal, not a model in the brain. Brain processes facilitate perception, but perception isn't a brain process. Perception is a way the animal, given its sensory and motor capacities, is related to the world. Enactive cognitive science shares this viewpoint.[62] Both ecological psychology and enactive cognitive science strive to create a science of perception that does not bifurcate nature.

The predictive processing theory distorts perceptual experience. We don't perceive predictions and error signals, even if our brain makes use of them. We perceive the world. How is the brain's guessing, and updating its guesses, equivalent to the concrete sense of shared presence we experience in perception? (In fact, the brain doesn't really make guesses or predictions; that's our way of talking, based on how we model the brain in particular contexts and under particular assumptions.) It's one thing to investigate how brain processes that we model in predictive terms may enable our perceptual experience of the world. It's another thing to declare that the content of our perceptual experience is one and the same thing as the content of a predictive model in the brain. The equivalence is unwarranted.

It surreptitiously substitutes a model of the content of brain processes for the content of perceptual experience. It also suffers from the fallacy of misplaced concreteness. It takes an abstract and idealized third-person model of the brain to be the concrete phenomenon of the first-person perceptual experience of the world. Finally, it confuses levels of description. Just as we don't perceive retinal images, we don't perceive predictions and error signals. There remains a fundamental discrepancy between predictive processing models and perceptual experience.[63]

But things get worse. The predictive processing theory is self-undermining when we apply it to perceptual knowledge, particularly our perceptual knowledge of the brain. If perception is "nothing more" than our brain's best guess about the hidden causes of its sensory inputs, then this must apply equally to our perception of the brain. The physical brain, which was supposed to be objective and outside the model, and the source of consciousness, now turns out to be a content inside the predictive model. It follows that we aren't really studying the physical brain when we do neuroscience, at least not directly. We're studying our "guesses" about a hidden, unobservable something that we call "the brain" from within the model. This point generalizes. Anything we perceive is just a "guess" made from inside the model. The model subsumes everything perceived to lie beyond it. The model swallows up its own physical basis.[64] That outside basis becomes just another hypothesis inside the model.

You could try to resist this conclusion by saying that every model needs a physical system that generates it, so there must be a physical brain outside the model, or at least an outside physical reality. But, strictly speaking, this doesn't logically follow. For all we know, the world outside the model could be entirely mental, as philosophical idealists think, or mathematical, as Platonists maintain. If there's no way to know what's outside the model, there's no way to establish what outside reality is and what concepts apply to it.

Our intention in making this argument isn't to dive into deep metaphysical questions about the relationship between appearances and things-as-they-are-in-themselves (to use Kant's terms). Instead, it's to point out that the predictive processing theory, when presented as a physicalist account of the generation of consciousness, undercuts itself, because it's not entitled to the notion of the outside-the-model, physical brain that it relies on. The failure to see this—that the theory postulates a physical source of consciousness that transgresses the theory and that the theory cannot account for—is another instance of the Blind Spot.

The predictive processing theory—extrapolated as a theory of perceptual experience or consciousness in general—undermines itself in another way. The theory makes it impossible to account for science, and hence for the theory itself. Science depends on shared perceptual knowledge in the scientific workshop and the life-world. But the predictive processing theory reduces shared perceptual knowledge to each person—or rather, each brain—coordinating its own private model with other private models. From the perspective of any given model, the existence of any other model is only a conjecture. Each perceiver relies only on private error signals to find its way about in the world. There is no credible way to account for scientific knowledge on the basis of this solipsistic framework.

The root of these intractable problems is thinking that we are our brains and that perceiving takes place inside them. This is like saying that a bird is its wings and that flying takes place inside them.[65] A bird needs wings to fly, but flight isn't inside the wings. Flying is a relation between the whole animal and its environment. We need brains to perceive, but perceiving isn't inside our brains. Perceiving is a relation between ourselves and the world. Once we collapse the distinction between ourselves and our brains, we create a false image of ourselves as trapped inside our heads, as if in a windowless room, having to guess what's going on outside, and working only with the discrepancies between what we guess and the sounds we hear coming from the other side of the wall.[66] But we're not inside our heads.[67] We're not reducible to our brains. The brain is an organ of perception, not the perceiver. The perceiver is the whole person or animal, geared into its world.

Integrated Information to the Rescue?

Another theory that purports to answer the question of how a physical system can generate consciousness is the integrated information theory of consciousness (IIT).[68] IIT theorists believe we should start from phenomenology, that is, from an examination of how consciousness appears experientially. Instead of trying to infer consciousness starting from physical systems, we should start from conscious experience and determine what kinds of properties physical systems must have to be conscious. Unlike phenomenologists, however, who patiently explore the ways we experience consciousness, IIT theorists jump right away to trying to identify essential experiential properties of consciousness, ones they believe are indubitable and true of every conceivable conscious experience.

Five essential properties are identified and serve as the "axioms" of the theory: (1) intrinsic existence (each experience exists from its own intrinsic or nonrelational perspective); (2) composition (each experience is composed of multiple phenomenal distinctions); (3) information (each experience is informative in the sense that it differs from other possible experiences); (4) integration (each experience is unified); and (5) exclusion (each experience is definite in content). These axioms are translated into "postulates" about the essential properties of the physical substrate of consciousness.

"Physical," in IIT, is understood abstractly as cause-effect power—how a system's parts mutually influence each other or how a system responds to all possible perturbations of its state. This notion of cause-effect power is used to translate the phenomenological axioms into physical postulates. For example, the axiom that each experience exists from its own intrinsic perspective is translated into the postulate that the physical substrate of experience must exist with its own intrinsic cause-effect power independent of extrinsic factors.

"Integrated information," also known as Phi or Φ, is defined at this causal level of interacting elements. Giulio Tononi, the inventor of the theory, describes integrated information as "the amount of information generated by a complex of elements, above and beyond the information generated by its parts," taking into account their causal and dynamical interactions.[69] Roughly speaking, integrated information is the information beyond that available from the sum of the system's parts and that can't be localized in the system's parts.

IIT's central thesis is that "consciousness is integrated information."[70] The level or amount of consciousness present in a physical system is said to correspond to the amount of integrated information (Φ) generated by the system's elements. The specific qualitative character of every conscious experience is said to be identical with an irreducible and intrinsic cause-effect structure (one that exists by itself independent of other things). This is described as a "form" in the space of informational relationships (possible cause-effect relations) among the system's elements. Finally, in relation to the brain, IIT specifies how the functional neural networks correlated with conscious states in the awake or dream states have a high amount of integrated information (Φ) compared to the brain states when consciousness is attenuated or absent in deep sleep or under anesthesia.[71] The theory also aims to specify how particular kinds of causal structures of the brain, understood in terms of integrated information, mirror (correspond one-to-one with) the essential properties of consciousness (as given in the theory's axioms).

IIT provides an interesting case study of the Blind Spot. Some aspects of the theory point beyond the Blind Spot, while others strongly reinforce it.

On the one hand, the concept and measures of integrated information have opened up new perspectives on complex systems beyond classical reductionism. For example, IIT theorists have shown how integrated-information measures of macrolevel system causation can supersede explanations of complex systems in terms of the microlevel causation of the system's elements.[72] In other words, integrated-information measures of the system as a whole can have greater explanatory power than explanations at the level of the causal interactions of the parts. Integrated-information concepts and measures have also been used to show how to delimit the boundary between an integrated unit and its environment and to illuminate aspects of biological organization.[73]

On the other hand, IIT's core thesis, that integrated information *is* consciousness, confuses the map with the territory. It's a striking case of surreptitious substitution—substituting integrated information for experience—and the fallacy of misplaced concreteness—treating integrated information, which is an abstraction from our knowledge of causal interdependencies, as if it were concretely real.

It's one thing to say that integrated information may be closely associated with aspects of consciousness and that we can use measures of integrated information as measures of the complexity of the neural networks associated with these aspects of consciousness. It's another thing to say that consciousness is identical with integrated information. IIT theorists believe this identity follows from their claim that there is an isomorphism (a one-to-one structural correspondence) between the components of the cause-effect structure of a physical system and the abstracted qualities of experience specified in IIT's axioms. But even if there were an isomorphism, rather than merely an analogy or formal resemblance, as actually is the case, the identity claim would not logically follow. An isomorphism is a relation that holds only for abstract objects distinguished solely on account of their structure, and isomorphism does not necessarily imply equality or identity. Conscious experience is a concrete phenomenon and not an abstract object. The abstract structure of consciousness does not exhaust what consciousness is.

IIT correctly sees that we have to begin with phenomenology when we investigate consciousness. But it immediately compromises this insight with surreptitious substitution. At the outset, the theory substitutes an axiomatic-deductive structure, which is necessarily based on abstraction and idealization, for the actual phenomena of consciousness. This substitution ignores the basic findings of phenomenology. As Husserl—who, remember, was a mathematician—long ago realized, the actual phenomena of consciousness are fluid and indeterminate (a point also made by James and Merleau-Ponty), and do not tend toward any ideal limit, unlike the entities of mathematical physics.[74] Bergson understood this crucial fact even earlier: recall the necessary fluidity and inexactitude of the experience of duration compared to clock time. Anyone who takes phenomenology seriously and knows its history would immediately question the appropriateness of trying to explain consciousness using an axiomatic-deductive structure. After all, scientific explanation hardly has to be axiomatic and deductive: biology doesn't explain life in terms of an axiomatic-deductive framework. Beginning with phenomenology means giving due attention to the actual phenomena of consciousness in their fluidity and indeterminateness instead of assuming at the outset that they can all be generated from axiomatic properties.

If we turn to IIT's specific axioms and postulates, which are supposed to underwrite the identification of consciousness with integrated information, we find that they are replete with problems. As philosopher Tim Bayne has shown, the axioms, despite being presented as self-evident, are either controversial or underspecified.[75] Let's review a few examples.

Axiom 1 (intrinsic existence) states that each experience exists from its own intrinsic or nonrelational perspective. This axiom depends on the problematic notion of an intrinsic property (a property something has all by itself without depending on anything else). The axiom amounts to a strong metaphysical claim—that any given experience has its own intrinsic character apart from its relations to anything else—a claim that is far from self-evident and that many theorists would reject. Embodied and enactive cognition theorists would argue that conscious states and their contents depend constitutively on their history and relations to the environment. Illusionists (discussed below) would say that conscious experiences may appear to exist intrinsically, but this is a cognitive illusion. And Madhyamaka (Middle Way) Buddhist philosophers would say that the very notion of anything existing intrinsically is incoherent.[76]

Axiom 2 (composition) states that each conscious experience is structured, that it's composed of phenomenal distinctions proper to it. This axiom is underspecified. Every theory recognizes that consciousness typically involves some degree or kind of phenomenal differentiation, but they disagree about how to describe it. For example, certain types of meditative states purport to be conscious and phenomenally unstructured.[77] Does this axiom rule them out? In that case, the axiom is disputable, not self-evident. Or does the axiom somehow allow for them (perhaps by counting them as cases of minimal structuration)? In that case, the axiom is underspecified.

Axiom 4 (integration) states that each conscious experience is unified. This axiom amounts to taking the side of the holists in the ancient and contemporary debates between phenomenal holists and phenomenal atomists.[78] Phenomenal holists maintain that one's overall conscious experience at a given time is an irreducible whole that can't be reduced to simpler phenomenal building blocks. Phenomenal atomists maintain that this impression of wholeness is an illusory mental construction and isn't real: experience can be exhaustively analyzed into elementary phenomenal atoms. Phenomenal holism may be true, but it's hardly self-evident. Furthermore, if the truth of holism or atomism is not discernible to introspection, then making holism axiomatic on introspective grounds won't work.

IIT's other two axioms are also either underspecified or not self-evident, but we don't need to go into the details.[79] The point to emphasize is that the identification of consciousness with integrated information rests on the IIT axioms and their translation into postulates about the properties of the physical substrate of consciousness. But these axioms are shaky, and the postulates aren't logically derived from them. What links the postulates to the axioms is resemblance or analogy, not logical derivation. For these reasons, IIT's central explanatory identification of consciousness with integrated information is unwarranted.[80]

IIT suffers from a deeper problem, one that takes us to the heart of the Blind Spot. The problem concerns its key concept of information. In our view, information, as defined in information theory, including integrated information theory, is fundamentally observer dependent and reflects the cognitive standpoint of an investigating agent.[81] Information is defined in terms of probability distributions, and probability distributions reflect an observer's or agent's uncertainty from a partial and limited point of view. In other words, states of the world, or states of a physical system, do not intrinsically carry information all by themselves; they carry information only from certain observational or cognitive perspectives. Information presupposes the existence of an observer, understood as a rational cognitive agent. From this perspective, building an account of consciousness out of information is backward or circular. The very concept of information presupposes consciousness in the form of the experiential knowledge of a rational cognitive agent. Information is a structural residue of a certain kind of experience—an agent's uncertainty regarding possible outcomes. Forgetting this truth about information is a case of the amnesia of experience.

Physicalism versus Panpsychism

It's time to step back and consider broader metaphysical ideas about consciousness in relation to the Blind Spot. Although the scientific theories of consciousness we've considered can be made consistent with different metaphysical frameworks, they're usually understood to be working within the framework of physicalism, though some panpsychists think IIT lends support to their position, while others criticize IIT for the same reason.[82] Physicalists believe that consciousness can be fully accounted for in terms of physical entities and processes, which they take to be fundamentally not mental. Life, mind, and consciousness, for the physicalist, are nothing over and above complex configurations of physical elements, which are ultimately reducible to atoms and elementary particles. Panpsychists, however, argue that the fundamental physical constituents of the universe are also mental. For panpsychists, mind or consciousness is as fundamental as quarks and photons. Although both physicalists and panpsychists think that human consciousness is complex and derives from simpler entities and processes, physicalists think those entities and processes are fundamentally physical and entirely nonmental, whereas panpsychists think they are fundamentally physical and fundamentally mental.

Our viewpoint is that physicalism and current versions of panpsychism take for granted the blind-spot conception of science and therefore operate within the ambit of blind-spot metaphysics. We'll start with physicalism and then take a look at panpsychism.

Physicalism, as a metaphysical thesis, is effectively useless. The reason is that "physical" is not well defined, and attempts to define it make physicalism either false, empty, or not naturalist in spirit, contrary to physicalism's original motivation.

Take the key physicalist claim that there's nothing but physical reality, or, to put it more precisely, that the universe or cosmos consists entirely and exclusively of physical entities and processes. To put it bluntly, this claim is either false or empty. On the one hand, if we define "physical" as what contemporary physics tells us is physical, then physicalism is false, since contemporary physics will inevitably be superseded by another physics with a new, refined, or enlarged conception of the physical. On the other hand, if we define "physical" according to some future, finished or ideal physics (an idea we think makes no sense[83]), then physicalism is empty, because we have next to no idea what such a future physics will look like.

The problem of physicalism being either false or empty is known as Hempel's dilemma, named after the twentieth-century philosopher of science Carl Gustav Hempel.[84] If physicalism is defined by way of physics today, then it is false, because contemporary physics is incomplete. But if physicalism is defined by way of reference to a future or ideal physics, then it is trivial or empty, because no one can say what a future physics contains. Perhaps, as a few physicists have argued, an ideal physics needs to contain even mental items or consciousness.[85] In that case, the dichotomy between the physical and the mental, on which traditional physicalism is based, would make no sense, assuming that "physical" means whatever physics must refer to. The conclusion of the dilemma is that we have no clear understanding of what it means for something to be physical, and hence no way to state physicalism as a meaningful thesis.

Faced with this quandary, some philosophers argue that we should define "physical" such that it rules out radical or strong emergentism (that consciousness is emergent from but irreducible to physical reality) and panpsychism (that mind is fundamental and exists everywhere, including at the microphysical scale). This move would give physicalism a definite content, but at the cost of trying to legislate in advance what "physical" can mean instead of leaving its meaning to be determined by physics.

We reject this move. It runs counter to the original spirit of physicalism, which is supposed to be scientific and naturalistic. Whatever "physical" means should be determined by physics, not armchair reflection. After all, the meaning of the term "physical" has changed dramatically since the seventeenth century. Matter was once thought to be inert, impenetrable, rigid, and subject only to deterministic and local interactions. Today we know that this is wrong in virtually all respects: we accept the existence of several fundamental forces, particles that have no mass (currently the photon and the eight gluons that mediate the strong nuclear force), quantum indeterminacy, and nonlocal relations. We should expect further dramatic changes in our concept of physical reality in the future. For these reasons, we can't simply legislate what the term "physical" can mean as a way to get out of Hempel's dilemma.

One panpsychist philosopher, Galen Strawson, has argued that physicalism, properly understood, actually entails panpsychism.[86] If that is the case, then physicalism allows, indeed even requires, that the fundamental physical entities also be fundamentally mental. This line of thought provides a further indication that it's very difficult to constrain the meaning of physicalism as a metaphysical thesis. Physicalism devolves to naturalism, a commitment to the scientific intelligibility of reality, in contrast to supernaturalism, the belief that there are phenomena or entities that aren't subject to the laws of nature. We can accept naturalism without grafting onto it the unhelpful metaphysics of physicalism.

Strawson's conception of physicalism brings us to panpsychism. He argues for panpsychism as follows:

1. Monism is true. (Reality is a concrete whole and isn't divided into fundamentally different kinds of things, such as immaterial minds and material bodies. In our terms, following Whitehead, nature is unified, not bifurcated.)

2. Experience is real. (Strawson calls illusionism, the thesis that experience is an illusion, "the great silliness."[87])

3. All real phenomena are physical. (Physicalism is true.)

4. Therefore, experience is physical.

5. The emergence of experience from nonexperiential entities and processes is impossible. (This is to deny radical or strong emergence. If radical emergence were true, experience would have to depend wholly on
nonexperiential entities and processes, so that experience would entirely
trace back to them, while being nonetheless novel from a purely physical standpoint. But we have no conceptual or scientific model for this
kind of emergence.)

6. So at least some elementary physical phenomena must also be experiential.

7. If at least some elementary physical phenomena are experiential, it's far
more likely that all of them are. (Otherwise there would be a radical heterogeneity or bifurcation in nature, contrary to the first point here.)

8. Therefore, all physical phenomena are also experiential phenomena,
which is to say that panpsychism is true.

The ingenuity of this argument is that it derives panpsychism by asserting physicalism and the reality of experience, on the one hand, and denying
the bifurcation of nature and radical or strong emergence for consciousness,
on the other. Physical reality must be fundamentally experiential given that
nature isn't bifurcated, everything is physical, experience is real, and the
emergence of experience from nonexperiential phenomena is impossible
(or unintelligible).

This argument is ambiguous with regard to the Blind Spot. It points
beyond the Blind Spot but ultimately stays within it.

The argument points beyond the Blind Spot by rejecting several crucial elements of blind-spot metaphysics, namely, the bifurcation of nature,
orthodox physicalism (physical reality is fundamentally non-mental), and
epiphenomenalism about experience. The argument proposes an undivided
conception of nature in which experience is irreducible. This opens the
door to recognizing the primacy of consciousness, the primacy of embodiment, and the strange loop between consciousness and the world.

Nonetheless, the argument remains hostage to the Blind Spot. The reasoning takes for granted the objectivist conception of physics (that physics
gives us access to reality apart from human experience) instead of recognizing that experience is ineliminably present everywhere in physics (the
structural invariants of physics are residues of experience) and therefore
that the physicalist picture of nature used to state the hard problem of

consciousness, to which panpsychism is supposed to be a solution, is faulty from the start. As a result, the panpsychist argument winds up injecting experience as an extra ingredient into an otherwise unchanged objectivist conception of physical reality. The upshot is to objectify experience, to make it an object out there in the physical world. This is paradoxical, because experience is precisely that which is not an object. Experience is the horizon within which any object or collection of objects is specifiable. The panpsychist argument inserts experience into nature without acknowledging that experience is a condition of possibility for the intelligibility of nature. Experience escapes objectification, but the panpsychist argument treats experience as if it were just a special kind of object. This is symptomatic of the Blind Spot.

The objectification of experience is evident in another panpsychist line of argument known as the "intrinsic nature" argument.[88] It goes as follows:

1. Physics reveals to us only the structural and relational properties of physical phenomena. (For example, physics tells us what an electron is only in terms of what it does, or how it's disposed to behave in relation to other entities; it doesn't tell us what the intrinsic nature of an electron is, what an electron is in and of itself.)

2. Relational properties are determined by intrinsic properties. (Relations depend on there being properties that the things standing in the relations have all by themselves.)

3. Certain configurations of physical phenomena (such as the brain) generate or constitute experiences.

4. So the intrinsic properties of physical phenomena must encompass this power to generate or constitute experiences.

5. Our own inner awareness reveals that phenomenal properties are intrinsic properties of our experiential states. (For example, we can tell by introspection that perceived blue is an intrinsic phenomenal property of the visual experience of seeing a clear blue sky.)

6. Phenomenal properties are the only intrinsic properties we know of.

7. So phenomenal properties must be intrinsic properties of physical phenomena, or at least of certain organized physical systems (such as the brain).

If we add to this argument the denial of radical emergence, and the claim that if phenomenal properties are intrinsic to at least some physical phenomena, they're likely to be intrinsic to all physical phenomena, then we arrive at panpsychism.

This argument objectifies experience by construing it as being explainable in terms of special phenomenal properties that constitute the intrinsic nature of the physical world. But experience can't be fully accounted for this way. Again, experience is the horizon within which any property is specifiable. It is the precondition for the manifestation of the world, not just something inside the world.

The argument has two other big problems. It depends on the unclear and problematic idea of an intrinsic property, and it rests on a mistaken conception of what we know about experience from our own inner awareness.

An intrinsic property is traditionally understood as a property that something would have even if it were the only thing in the universe or the only thing in existence. Does that idea even make sense? Not if you think that something is what it is only by virtue of its belonging to a web of relations. Why not say that relations determine the occupants of the relations, after the fashion of relational quantum mechanics (see chapter 4)? Or that relations and occupants are mutually interdependent? Madhyamaka (Middle Way) Buddhist philosophers, over centuries of discussion with their Indian philosophical interlocutors, have given compelling reasons to reject the intelligibility and existence of intrinsic properties.[89] Their arguments have inspired analytical philosophers and quantum physicists to maintain the primacy of relations over entities with intrinsic properties.[90] In any case, the above argument turns on the metaphysical concept of an intrinsic property, a concept that is very difficult to make sense of.

In addition, it's hardly the case that we know from our own inner awareness that the qualities of our experience are intrinsic properties of our experiential states. Can you tell just by introspecting that the blue you see when you look at the clear blue sky is an intrinsic property of your experiential state, where that means that it isn't constituted out of relations involving your body, environment, and personal history? Certainly not. Indeed, careful introspection suggests precisely the opposite. It suggests that the qualities we experience are constituted by relations— relations that come from having a body, being situated in an environment, and having a history. The blue of the sky doesn't present as an intrinsic mental property but as an experiential property of the world, one that is saturated by the whole visual context, the body's responsiveness, and the sedimented memories of similar past experiences. These myriad relations are inexhaustible, certainly to introspection, in their breadth and depth. More generally, inner awareness doesn't give us access to the full nature of our experiential states, and anything that seems to be intrinsic is likely to turn out to be relational.

The problem with current versions of panpsychism is that they accept the blind-spot conception of science and the physicalist picture of reality, but then sprinkle consciousness over physical reality everywhere like pixie dust. Although this gives consciousness a kind of primacy, it's not the right kind to supplant the Blind Spot. The primacy of consciousness as a special kind of building block is not the primacy of consciousness as that within which anything appears to us, including whatever we may take as building blocks. Unless we recognize the second kind of primacy, the primacy of horizonal consciousness, including how it necessarily enfolds the primacy of embodiment, we won't be able to recognize the strange loop that obviates the Blind Spot. At the end of the day, current versions of panpsychism amount to a stopgap for dealing with the problem of consciousness within the confines of the Blind Spot.

Just an Illusion

Some theorists have suggested that if the search for a theory of consciousness leads all the way to panpsychism, then we should question whether consciousness really exists.[91] This idea brings us to illusionism, the last main way of trying to deal with consciousness within the Blind Spot.

Illusionists say that consciousness is an illusion, that it isn't what it seems to be.[92] The illusion they're concerned with is a kind of cognitive illusion, a mistaken impression that comes from a false way of thinking. Consciousness seems to possess intrinsic, subjective, qualitative properties, so-called qualia or phenomenal properties. Illusionists say there are no such properties.

Suppose you burn your finger on a hot pan. If you think in terms of qualia—which is a philosophical construct, not an everyday notion—you'll think that you're encountering a particular phenomenal property, the property of searing painfulness. You'll think that property is inherently subjective, private, and qualitative and belongs intrinsically to your pain state. Illusionists deny that experience really has such properties. They don't deny the experience of pain; they deny that there are pain qualia. They deny that the experience of pain consists in being immediately acquainted with intrinsic properties that are inherently subjective, private, and qualitative. Instead, they maintain that the experience of pain consists in being in a certain physical and informational state generated by the brain. Introspection represents that brain state in a simplified way—the way a desktop folder icon on a computer screen represents the location and content of data in a simplified way. Introspection depicts the underlying brain state as if it had uniquely qualitative properties, but it doesn't really. The brain state consists entirely of physical, informational, and functional properties, which are describable without mentioning anything inherently qualitative or phenomenal. The impression of subjectivity and privacy comes from the brain's being self-monitoring and self-representing and thereby having a unique (but limited) kind of access to its own states.

For the illusionist, consciousness is not a special medium for phenomenal properties. It is a kind of complex cognitive and informational process. To be conscious is to use information flexibly by thinking about it, paying attention to it, remembering it, reporting it, and so on. Consciousness may seem to be populated by phenomenal properties, but that is an illusion. The illusion comes from applying a mistaken concept, the concept of qualia or phenomenal properties, to how introspection simplifies and misrepresents the brain states underlying experience.

Illusionism is often derided by its opponents because it seems to deny something undeniable—consciousness in the sense of direct experience or awareness with a phenomenal character.[93] That may be so, but illusionism contains an important insight. The insight is that qualia are a theoretical construct, not something you can read off of experience, and that construct is fraught with problems. "Qualia" doesn't just mean how things look or feel, or generally seem when we experience them, though philosophers often use the word in that anodyne way. *Qualia* is a technical term that refers to the purported intrinsic, qualitative, or phenomenal properties that are thought to determine how things seem when we experience them. Qualia, so understood, are theoretical posits, not data. They require us to think that experience has qualitative characteristics whose identity and nature don't depend on the relations in which they figure, particularly the relations between the body and the environment and the relations of experiences to one another.

Illusionists aren't the only ones who deny that there are qualia. Phenomenologists do too. They think that the concept of qualia distorts lived experience. As Merleau-Ponty writes, "The pure *quale* would only be given to us if the world were a spectacle and one's own body a mechanism with which an impartial mind could become acquainted. Sensing, however, invests the quality with a living value, grasps it first in its signification for us, for this weighty mass that is our body, and as a result sensing always includes a reference to the body."[94] By "pure *quale*" Merleau-Ponty means a "raw feel," a phenomenal property having an intrinsic, qualitative nature, apart from the bodily and environmental relations in which it figures and with no reference to anything beyond itself. This idea reflects the bifurcation of nature, the division of nature into physical properties (primary qualities) and mental qualities (secondary qualities), or in Merleau-Ponty's terms, the body as a mechanism, which responds to physical stimuli, and an impartial mind, which registers them as qualia. Merleau-Ponty rejects this picture. He replaces it with a conception of sense qualities as constituted by the living body's meaningful relations to the environment ("this red would literally not be the same if it were not the 'wooly red' of a carpet"[95]) and to the living body's affective and motivational tendencies (this red would literally not be the same if it were not the mouth-watering red of a strawberry).

There's a big difference, however, between phenomenology and illusionism. Phenomenologists (particularly those who follow Merleau-Ponty) recognize the primacy of consciousness, the primacy of embodiment, and the strange loop, whereas illusionists think that consciousness is an illusion that can be accounted for in entirely objective physical terms. The difference is day and night.

The problem with illusionism isn't its denial of qualia. The problem is that it confuses the misbegotten concept of qualia with direct experience. It confuses the incontrovertible fact of direct experience with an intellectual theory that distorts direct experience. It's one thing to dismiss the concept of qualia as a theorist's fiction. It's another to say that direct experience is an illusion created by the brain. That statement is self-undermining because we have no knowledge of the brain apart from our direct experience of it. Take away direct experience, and we have no knowledge of anything.

Illusionists are typically physicalists.[96] They assume that only physical reality exists, that science provides access to objective reality outside of experience, and that physical reality is fundamentally nonexperiential. It's impossible to account for experience in this framework. The reason isn't the resistance of qualia to physicalist explanation. The reason is that direct experience is a precondition and essential ingredient of science. Illusionists think they can explain experience exhaustively in terms of information-processing and functionalist models of the brain. That's impossible. These models are structural residues of experience. They presuppose experience as a necessary condition of their own possibility and intelligibility. Thinking that one can explain direct experience exhaustively without remainder in terms of its structural residues is nonsensical. Physicalist illusionism remains entrenched in the Blind Spot.

A Science of Consciousness in Which Experience Really Matters

Three decades ago, when David Chalmers first coined the term "the hard problem of consciousness" and called attention to its importance, a few scientists made the bold suggestion that the intractability of that problem within what we are calling the Blind Spot means that we need to reframe science in order to investigate consciousness.[97] In two independent papers, published in the same issue of *Journal of Consciousness Studies* in 1996, astrophysicist Piet Hut and cognitive psychologist Roger Shepard, and neuroscientist Francisco Varela, made the case for a major overhaul of the science of consciousness based on recognizing the primacy of experience.[98] They pointed out that we inescapably use consciousness to study consciousness. So unless we recover from the amnesia of experience and restore the primacy of experience in our conception of science, we'll never be able to put the science of consciousness on a proper footing. The problem arises when we suppress the primacy of consciousness and try to assimilate consciousness to its structural residues. But simply recognizing the inescapable need to use consciousness to study it is only a first step. We also need better methods for using consciousness this way. This idea of a methodological transformation of the problem of consciousness into a new scientific research program lay at the heart of what these scientists proposed. We end this chapter by revisiting their ideas for a science of consciousness in which experience really matters.

Hut and Shepard's proposal has two steps, which they call turning the hard problem upside down and then turning it sideways. The first step is to flip the problem of consciousness on its head by starting from direct experience instead of from the physical world, posited as independent of experience. This step works as a corrective to the objectivist bias operating in the Blind Spot. It inverts our perspective from a third-person one to a first-person one. The second step is to turn the dial of experience away from a first-person setting to a second-person one. Experience is inherently intersubjective and shareable. Science would be impossible if this were not the case.

The first step, turning the hard problem upside down, means asking how what we understand to be physical objects arises out of experience, instead of asking how experiences arise out of physical objects (such as the brain). For example, instead of asking how to account for experiential time, particularly the flowing now of duration, on the basis of physical time, in which there is no flowing now, we ask how to account for physical time as an abstraction from experiential time (see chapters 3 and 5). (This was actually Husserl's project in his 1905 lectures on the consciousness of time and part of Whitehead's project in *The Concept of Nature.*[99]) This inversion of the order of explanatory priority, from a taken-for-granted, objective physical world, to experience as the basis for any scientific representation of the world, signals the adoption of a phenomenological stance in which we recognize the primacy of lived experience.

The second step, turning the hard problem sideways, means realizing that I-and-you experience already permeates I-experience. Your ability to reflect on your own experience comes from your having internalized outside perspectives on yourself as you mentally developed in infancy and childhood. In the language of cognitive science, this kind of metacognition (reflection on your own experience) is a special case of social cognition (cognition about self and other). In addition, being able to say anything about your experience depends on having a language you share with others. So the first-person singular perspective, which initially ensues when we flip the hard problem upside down, cannot be absolutely privileged. Instead, we need to turn the experiential perspective around to make its I-and-you structure explicit.

Taking these two steps, which, in our terms, consist in recognizing the primacy of consciousness and the primacy of embodiment, changes how we think about the problem of consciousness. The problem for neuroscience can no longer be stated as how the brain generates consciousness. Instead, the problem is how the brain as a perceptual object within consciousness relates to the brain as part of the embodied conditions for consciousness, including the perceptual experience of the brain as a scientific object. The problem is to relate the primacy of consciousness to the primacy of embodiment without privileging one over the other or collapsing one onto the other. The situation is inherently reflexive and self-referential: instead of simply regarding experience as something that arises from the brain, we also have to regard the brain as something that arises within experience. We are in the strange loop.

The strange loop's presence in neuroscience becomes explicit in the case of real-time neurofeedback. Here the experimental participant has access to real-time online signals from their own brain. Suppose your task is to pay attention to your breath while you're given an image of your own brain activity corresponding to your degree of attentiveness or distraction (an experiment that actually has been done).[100] As Hut and Shepard write, "Ultimately, we encounter this curious circle: part of the 'lighting up' in the brain image I experience may represent the very neuronal activity that corresponds to my experiencing the brain image of that same activation. This may lead to Gödelian paradoxes."[101]

Varela's step was to propose a neuroscience research program, called "neurophenomenology," based on braiding together first-person accounts of consciousness with third-person accounts of the brain within the I-and-you experiential realm. Concretely, this means establishing "mutual constraints" between the two kinds of accounts. Phenomenology and neuroscience become equal partners in an investigation that proceeds by creating new experiences in a new kind of scientific workshop, the neurophenomenological laboratory. First-person experiential methods for refining attention and awareness (such as meditation), together with second-person qualitative methods for interviewing individuals about the fine texture of their experience, are used to produce new experiences, which serve as touchstones for advancing phenomenology. This new phenomenology guides investigations of the brain, while investigations of the brain are used to motivate and refine phenomenology in a mutually illuminating loop.

Neurophenomenology is based on the idea that a deep investigation of consciousness requires working with individuals who are skillful at the kind of meta-awareness of experience we described at the beginning of this chapter. Since training these skills is emphasized in Asian meditative traditions, neurophenomenology draws from these traditions to enlarge and enrich phenomenology. The working assumption is that individuals who can generate and sustain "mindful meta-awareness" can make qualitatively and temporally fine-grained reports about their moment-to-moment experience, and these reports can be used to reveal patterns of activity in the brain and the rest of the body that would otherwise be missed.[102] This approach also involves using precise qualitative methods of interviewing individuals about fine-grained characteristics of their tacit experience.[103] Phenomenological reports produced in this careful way can help to pick out and ascribe meaning to previously unnoticed neurophysiological patterns, such as brain-heart interactions, while such new patterns can also be used to gain new insight into experience. This is what Varela meant by "mutual constraints" between phenomenology and neuroscience.

Neurophenomenology has led to a wide range of fruitful investigations in the neuroscience of consciousness and is a growing research area.[104] We call attention to it here, because it represents probably the strongest effort so far to envision a neuroscience of consciousness beyond the Blind Spot.[105]

IV The Planet

9 Earth

The Blind Spot on a Planetary Scale

In the first quarter of the twenty-first century, two basic facts about humanity's ten-thousand-year-old global project of civilization have become clear. The first is that global warming, driven by the industrial activity of the richest nations, is changing Earth's climate in ways that will severely stress that project. The second is that the global scale of habitat destruction associated with industrial activity can drive global pandemics that also pose severe threats to that project. Thus, the coupled natural and technological systems upon which almost 8 billion people depend for life can no longer be assumed to sustain their operation into the indefinite future.

Both realizations—of the risks of climate change and of global pandemics—have occurred within the rise of a new understanding of Earth itself. Beginning in the early twentieth century, a new scientific vision of the planet was slowly and painstakingly assembled. That vision saw Earth as more than simply a spherical rock harboring a thin and unimportant scruff of life on its outer edges. Instead, a suite of new conceptual tools created a perspective that saw the planet in terms of a set of strongly coupled systems. One of these systems was the totality of life (the biosphere), which, in this new perspective, emerged as an essential player in planetary evolution. Earth was no longer the inert background stage that allowed for the drama of life to unfold; life itself fed back into geological and atmospheric cycles, generating an ever-changing, mutually contingent, complex interacting dynamic state. In short, planetary changes lead to life changes, and life changes lead to planetary changes, in an amplification of the autonomy and agency of living systems to planetary scales.

The new understanding of Earth is important for our efforts to look beyond the Blind Spot. Over the past two hundred years, we've built an entire global culture based on treating Earth according to the strictures of the Blind Spot. The consequences of this construction have now become painfully apparent as humanity pushes Earth into a new planetary state and human-shaped epoch called the Anthropocene. Climate change and global pandemics are its most apparent manifestations. The Anthropocene signifies a new version of the planet that is likely to be far less hospitable to our project of civilization. At its worst, the changes we see beginning to occur may lead to that project's collapse.

The Anthropocene is a massive manifestation of the Blind Spot. It's the result of one particular and very recent version of the human civilizational project: the originally modern European, and now transnational, scientific project of objectifying the world through scientific materialism.

Ironically, the Anthropocene is also the starting point for a new scientific perspective on Earth that can help us move beyond the Blind Spot. Scientists now understand that the planet must be considered as an integrated whole—a coupled set of systems constituting the atmosphere, hydrosphere (oceans), cryosphere (ice), lithosphere (rock), and biosphere. The inclusion of the biosphere represents a critical transition in scientific thinking. Life is understood as a key player in the entire evolutionary history of Earth.

The constellation of new scientific disciplines that form the foundation for the current understanding of Earth and the biosphere are significant in relation to the Blind Spot. Network theory, cybernetics, dynamical systems theory, chaos theory: taken separately, each of these fields challenges different aspects of the Blind Spot's metaphysical assumptions about life, the world, and experience. Taken together in what is called complex systems theory, they represent the emergence of a new way of seeing how science functions, what it describes, and, most important, how it relates to human experience. With the crisis of climate change and the sustainability of the project of civilization in mind, we will describe the emergence of this new conception of Earth (and all planets), and the new sciences on which that view depends.

As we describe the sciences responsible for this vision of Earth and its biosphere, we must remember that the Blind Spot has always been about more than just science. That's because science has always been about more than just a search for reality and truth. From Francis Bacon on, developing the methods of inquiry we now call science was always aimed at the control of nature as much as at revealing reality and truth.[1] For Bacon, this control was intended to better the human condition by reducing sickness, famine, and our fragility from the vagaries of storms, floods, and earthquakes. Of course, Bacon was right about this. The spectacular success of science did provide greater control over natural events and processes than ever before. But the development of scientific and technological capacities emerged together (via the scientific workshop), and with them came an equally profound enlargement of human productive capacities. The age of science and the age of industry were born together, and each depended on the other.

The continued growth and success of science through the nineteenth and twentieth centuries, regardless of whether it happened under a capitalist, socialist, or communist banner, was inextricably linked to the rise of industrial-scale capacities of resource extraction, transformation, and consumption. Energy was harvested and entropy generated on scales that would soon alter the function of the coupled Earth systems. The economic, social, and cultural systems that emerged during this period were not separate from the science and technology on which they depended. The reverse is also true: the emergence of science's institutions, capacities, and outlook were not separate from the emerging culture that supported it and gave it broad power and reach.

As a result, the basic assumptions of the Blind Spot became entrenched in the economic and social systems that came to dominate the period of rapid industrialization, sometimes called the "Great Acceleration." Given that those systems must be implicated in the continuing failure to respond to the crisis of climate change, we also briefly explore the braiding of science, industry, economies, and culture during the rise of the Blind Spot as a dominant metaphysics of science. In addition, we attempt to outline how the emerging unifying view of Earth and life as a coupled complex system may help to find new ways of reimagining the relationship of the human project of civilization to the rest of the planet.

The Living Planet: From Geology to Biosphere to Gaia

By the end of the nineteenth century, the scientific disciplines and their boundaries as we know them today had become well established. At any university, students would find that life was studied in the biology department, chemicals in the chemistry department, and matter in motion in the physics department. If you had questions about the Earth, you would have been directed to the geology department. But for one of that era's brightest young geologists, Vladimir Vernadsky, the rigid disciplinary boundaries separating his science from the others made little sense.

Vernadsky began his scientific work in the chemical study of minerals. Traveling across Europe in the late 1800s, he was keen to apply the most modern methods of physics to the study of rocks, hoping to bring precision tools to bear on questions about the planet's history. Vernadsky, however, was always more than a specialist. Although he began his career studying minerals, his attention was always focused on the whole that emerges from the narrower stories scientists unlock from investigating the parts.

This focus on the integration of different disciplinary domains led Vernadsky to build a new field called geochemistry. It unfolded Earth's history by examining the microscopic composition of its physical constituents. But Vernadsky also saw that it wasn't just the physical and chemical components of the Earth that formed a dynamical whole. Realizing that biology must also be brought into the planet's story at a fundamental level, he created a second new domain called biogeochemistry.

Vernadsky's broad perspective concerned the evolution of life as much as the evolution of rocks. While each species was shaped by its particular local niche, it was also shaped by the activity of life as a whole on the planet. The evolutionary effects ran both ways. Earth too was shaped by the totality of life. As Vernadsky put it, "An organism is involved with the environment to which it is not only adapted but which is adapted to it as well."[2] A planet's life history is deeply enmeshed with life's history in that planet.

This attention to both microscopic and macroscopic views led Vernadsky to his most important addition to the language of life in the context of its planetary host. Building from discussions with the Swiss geologist Eduard Suess, Vernadsky proposed that the study of Earth would not be complete without understanding the central role of life as a planetary force. Earth, in his view, cannot be truly understood without understanding the dynamics of its biosphere. The addition of "biosphere" into the lexicon of Earth science was a recognition that life is a planetary power as important as volcanoes and tides. Life is an agent, shaping the complex multibillion-year history of the world. Vernadsky wrote in 1926:

Activated by radiation, the matter of the biosphere collects and redistributes solar energy and converts it ultimately into free energy capable of doing work on Earth. A new character is imparted to the planet by this powerful cosmic force. The radiations that pour upon the Earth cause the biosphere to take on properties unknown to lifeless planetary surfaces, and thus transform the face of the Earth. Activated by radiation, the matter of the biosphere collects and redistributes solar energy and converts it ultimately into free energy capable of doing work on Earth.[3]

And, a few pages later:

Life exists only in the biosphere; organisms are found only in the thin outer layers of the Earth's crust, and are always separated from the surrounding inert matter by a clear and firm boundary. Living organisms have never been produced by inert matter. In its life, its death, and its decomposition an organism circulates its atoms through the biosphere over and over again, but living matter is always generated from life itself.[4]

For Vernadsky, the biosphere was an extended region reaching from below the Earth's crust (the lithosphere) to the edge of the atmosphere. Within this shell—now called the "critical region"—the action of life dramatically changes flows of matter and energy: "The evolution of living matter continuously penetrates the whole biosphere . . . influencing also its inert natural bodies. That is why we can and must speak about the evolutionary process of the biosphere in its totality. . . . The evolution of species turns into the evolution of biosphere."[5]

Vernadsky saw that the world-shaping powers of life were both ancient and ongoing. "Adjusting gradually and slowly, life seized the biosphere," he wrote, and "this process is not yet over."[6]

Vernadsky's conception of the biosphere as a substantial, and even equal, player in the evolution of Earth was slow to gain acceptance. It was not until the idea was given a more radical caste that its importance was fully recognized.

In the early 1960s, independent scientist and polymath James Lovelock was hired by a fledgling NASA to help design potential experiments that could search for life on Mars. Reflecting on the difference between searching for "life as we know it" compared to designing more agnostic search techniques, Lovelock recognized that a robust biosphere would change its host planet's atmosphere. Two billion years ago, it was Earth's microbial life that created its high levels of atmospheric oxygen. The presence of this oxygen drives the Earth's atmosphere out of chemical equilibrium, forcing it into a state that a dead world (one without life) could not achieve. What was remarkable for Lovelock was that life had somehow managed to maintain the atmospheric state (i.e., the oxygen concentration) at a roughly steady level for hundreds of millions of years. Contemplating the chemical pathways through which life could maintain this forcing, Lovelock was struck by the idea that there must exist feedback mechanisms between life and the planet as a whole that maintain the atmosphere's disequilibrium state for extended periods of time. Just as an organism can maintain its internal conditions, such as its salinity or temperature, Lovelock thought that such feedbacks coupled the living and nonliving components of the planet together in ways that could maintain steady disequilibrium conditions. Going further, Lovelock imagined that life could evolve feedbacks that would maintain the planet in conditions that were favorable to itself.

Lovelock originally wanted to call this idea "Self-regulating Earth System Theory."[7] The essential terms are *self-regulation* and *system*. Lovelock was explicitly conceiving of the planet as a collection of systems, associated with the air, water, land, and life, that were strongly connected or "coupled." These systems could not be treated separately because each depended for its properties and behaviors on the others. This coupling implied a co-evolution in the planet's systems. The biosphere, atmosphere, hydrosphere, cryosphere, and lithosphere evolved together, leading to a global regulation of global properties, such as the Earth's average temperature. This self-regulation, according to Lovelock, served to maintain global properties in a domain favorable to life, even as external conditions, such as the slowly evolving sun, changed in ways that should make the global properties less favorable to life.

This systems perspective, with its emphasis on self-regulation and control, did not come to Lovelock in a vacuum. It was part of a growing interest in what was called cybernetics—the study of communication and control in biological and artificial systems. Cybernetics had roots in the study of self-regulating systems. It emphasized "circular causality" or recursive causal relations among the processes of a system rather than linear cause-effect ones.[8] For example, the biochemical processes that make up a cell recursively produce themselves over time, so that processes and products mutually determine each other (see chapter 6). As we will see, this kind of systems-based thinking would come to represent a novel approach to the world and to science.

Of course, Lovelock's new idea was not called "Self-regulating Earth System Theory." Instead, he named it the "Gaia hypothesis," and later the "Gaia theory."[9] Gaia was the Greek goddess of the Earth, and the more evocative name was suggested to Lovelock by his neighbor, novelist William Golding, author of *Lord of the Flies*.

Gaia theory was incomplete, however, without a clearer conception of where the feedbacks necessary for self-regulation originated. This was provided by Lynn Margulis, the brilliant and iconoclastic biologist who joined Lovelock as co-creator of the full Gaia theory. Margulis, whose work focused on the microbial foundations of the biosphere, had become famous for establishing the pivotal role of cooperation, rather than simply competition, in the evolution of life. She had amassed overwhelming evidence that cooperation, in the form of symbiosis, had driven the formation of the all-important eukaryotic cells from simpler prokaryotic partners.[10] Eukaryotic cells with their nuclei and organelles, such as mitochondria, are the basis for all complex cellular forms of life on Earth (plants, animals, and fungi).

The vast webs of microbial life and their evolution formed the biospheric basis for Lovelock and Margulis's enlarged version of Gaia theory. Recursive loops between microbial populations and the geophysical/geochemical environment composed the self-regulating mechanisms by which the full coupled system was proposed to regulate itself. One example that Lovelock and collaborators proposed is the production of dimethylsulphide (DMS) by phytoplankton.[11] DMS is a chemical associated with cloud formation, and phytoplankton are ocean-born photosynthesizing microbes. The logic of the feedback begins with a small increase (a perturbation) in solar energy hitting the planet's surface that would, by itself, act to warm the planet. But more sunlight will also lead to an increase of the phytoplankton population as they feed on solar photons. This leads to the emission of more DMS. But since DMS seeds cloud formation, the end result will be a cloudier planet that reflects more sunlight back into space. In this way, the feedback loop is completed, and the planetary temperature is brought back down. Note that what we just described is a negative feedback between microbial life and the geophysical/geochemical environment. An upward perturbation to the system gets deamplified, returning the system to its original state, that is, a self-regulating feedback loop is formed. Systems views also include the possibility of positive feedbacks that amplify perturbations. In addition, moving beyond the classical engineering idea of feedback, systems theory developed models of autonomous networks, such as the autopoietic cell, that produce their own operational processes and constraints (as discussed in chapter 6). For Margulis, autopoiesis became the crucial concept for conceptualizing how Gaia (Earth's coupled systems) could produce and regulate themselves.[12]

A dense, evolving web of recursive loops, grounded in microbial activity, represented the fullest vision of Lovelock and Margulis's Gaia theory, which they would spend decades articulating and refining. The theory was controversial when it was first proposed. Some scientists thought that it introduced a teleological principle into evolution. But this criticism arose from not taking into account the cybernetic logic of the theory, according to which the maintenance of the Earth's coupled systems in conditions conducive to life happens through recursive causal loops and not from striving to realize a goal. Others argued that there was no way for such a self-regulating, planetary-scale system to arise through natural selection. But this criticism was also misdirected because the original Gaia hypothesis, and the later Gaia theory, required only the lawful emergence of global properties through dense webs of recursive causation in an individual complex system, not natural selection acting on a reproductive population.[13] There do remain important questions concerning the evolution and efficacy of biospheric feedbacks for producing a full planetary homeostasis.[14] Nevertheless, recent work has pointed to evolutionary mechanisms that may select for the global-scale negative feedbacks that could maintain such a system.[15]

What matters for us, however, is not whether Lovelock and Margulis were formally correct in positing that life on Earth could establish full self-regulation of global conditions. Whether such a complete form of autonomous planetary self-regulation is, or can be, achieved remains an open question. Instead, it's the emerging alternatives to Blind Spot narratives of science embodied by Vernadsky, and Lovelock and Margulis, that we wish to focus on. This focus is particularly warranted given what happened next to Lovelock and Margulis' Gaian idea and the role it played in recognizing the Anthropocene and the existential crisis of climate change.

Earth Systems Science, Climate Change, and the Anthropocene

Gaia theory, with its emphasis on the strong coupling between life as a system (the biosphere) with the other abiotic Earth systems, represented a crucial step in the emergence of what is now called Earth system science (ESS). Defined as "a transdisciplinary endeavor aimed at understanding the structure and functioning of the Earth as a complex, adaptive system," ESS is now the foundation for understanding Earth's evolution as well as the evolution of planets in general.[16] A review of the history of ESS, led by climate scientist Will Steffen, puts it this way:

> For tens of thousands of years, indigenous cultures around the world have recognized cycles and systems in the environment, and that humans are an integral part of these. However, it was only in the early 20th century that contemporary systems thinking was applied to the Earth, initiating the emergence of Earth System Science (ESS). Building on the recognition that life exerts a strong influence on the Earth's chemical and physical environment, ESS originated in a Cold War context with the rise of environmental and complex system sciences.[17]

Tracking the history of ESS, Steffen and collaborators cite both Vernadsky's works on the biosphere in 1926 and the papers detailing Gaia theory in the 1970s as crucial publications setting the stage for ESS.

The next milestone they cite was the publication of the first Bretherton diagram by meteorologist Francis P. Bretherton in 1986. Embodying a systems perspective, a Bretherton diagram represents Earth's dynamics as a series of boxes (representing different systems and subsystems) linked by arrows that detail their couplings. Looking like an electronic circuit, a Bretherton diagram strikes the viewer as a complex tangle of causes and effects. That apparent complexity, however, captures the essence of what was being recognized about Earth: it must be treated as a whole emerging from what only appear to be separate components (each treated by their separate academic disciplines). Bretherton diagrams for different aspects of Earth's behavior soon proliferated as researchers attempted to dive into the dynamics of ESS's various subcomponents and subsystems (such as glacial evolution, CO_2 cycles, and ocean currents).

ESS, with its emphasis on systems-level thinking and emergence, represented a different approach to science than that of Blind Spot reductionism. Although the work of ESS researchers sometimes uses reduction as a method (looking at systems in terms of their components), it is not reductionist in its view of systems. It recognizes that the emergent system level is what matters most for understanding behavior. Hence no single existing discipline can lead the way to understanding the planet. Earth, as a "living planet" with a biosphere as one of its core systems, is not the province of the geology, biology, chemistry, and physics departments taken singly or even added together. Instead, something entirely new is required. Earth systems science isn't just an interdisciplinary endeavor; it is transdisciplinary. Interdisciplinary work means researchers from different disciplines cross boundaries to work together. But in transdisciplinary studies those boundaries fade, as entirely new perspectives and approaches are constructed.

It is noteworthy that the military's initial funding of climate-related science in the 1950s and 1960s gave the nascent field the jump-start it required to begin building a global view of global climate dynamics. Camp Century, a nuclear-powered military base established high on the Greenland ice sheet, was one potent example of these projects. It was there that the first deep ice core drilling operations began. By recovering ancient water from almost a mile below the ice sheet's surface, the planet's climate history over the last twelve thousand years was recovered. Recorded in those ice cores was a detailed view of the end of the last ice age and the surprisingly rapid climate change that accompanied it. The emerging recognition that Earth's global climate conditions could change dramatically on timescales of centuries or even decades would come to haunt the emerging field.

These efforts produced early fruit in the understanding that collective human activity had the capacity to affect planetary dynamics. The recognition that global warming could be driven by fossil fuel use dates back to 1908, when Swedish physicist and chemist Svante Arrhenius predicted average temperature increases for the planet due to coal consumption.[18] Until the 1950s, however, there remained unresolved questions about the possibility of such warming. In particular, some researchers believed the ocean had the capacity to absorb and store any additional CO_2 we added to the atmosphere. In addition, there were no high-quality measurements of atmospheric carbon dioxide that could be used to track global atmospheric inputs due to human activity. But in 1956, geochemist and oceanographer Charles David Keeling established a CO_2 "observatory" atop the Mona Kea volcano. Mona Kea was high enough to allow the samples taken there to represent a true measure of average global atmospheric concentrations. After just a few years, the steady increase in CO_2 due to fossil fuel burning was apparent in Keeling's data. In addition, Keeling's collaborators, Rodger Revelle and Hans Suess, soon showed that while the oceans had provided significant storage of CO_2, they would be unlikely to continue to act as a sink in the future as carbon dioxide levels increased. Their landmark 1957 paper on ocean uptake of CO_2 included a famous recognition that "human beings are now carrying out a large-scale geophysical experiment of a kind that could not have happened in the past nor be reproduced in the future."[19]

The work of Revelle, Keeling, and their collaborators did not remain cloistered in the academy. Working its way through the government's science advisory networks, the potential threat of climate change made it all the way to the office of the US president. On November 5, 1965, Lyndon Johnson raised the issue of global warming to a joint session of congress where he said, "This generation has altered the composition of the atmosphere on a global scale through . . . a steady increase in carbon dioxide from the burning of fossil fuels."[20] Looking back, it is remarkable that the dangers of a changing climate were known at the highest levels of US policymaking as far back as the 1960s. Of course, this concern was not acted on, and it never reached broad public consciousness during this time.

Concern for the human impacts on climate among scientists, however, did rise through the 1980s. During the early years of this decade, an intense effort was begun to predict and measure the "signal" of global warming, meaning the rise of global average temperatures above the inherently noisy background of year-to-year variations. Measuring this signal and articulating the broader consequences of a changing planetary climate spurred the international, transdisciplinary growth of Earth systems science as a scientific field. A host of new international scientific collaborations were formed with the explicit goal of articulating the reach and methods of the newly emerging ESS. These included an alphabet soup of programs and organizations: the World Climate Research Program (WCRP) begun in 1980, the Geosphere-Biosphere Program (IGBP) launched in 1986, a global program to study biodiversity called DIVERSITAS that was initiated in 1991, the International Human Dimensions Program (IHDP) on Global Environmental Change created in 1996, and, most important, the now famous Intergovernmental Panel on Climate Change (IPCC) created in 1988.

By the late 1980s and early 1990s, global warming finally began appearing in measurements of globally averaged temperature. The "signal" was climbing above the noise. At a congressional testimony on a blistering day in July 1988, climate scientist James Hansen proclaimed that "climate change is here." Hansen's testimony, which made news around the world, represented the introduction of the greenhouse effect, global warming, and climate change into widescale public consciousness and debate.

In the decades that followed, the scientific consensus that anthropogenic climate change was occurring only strengthened. The global development of ESS as a scientific endeavor made explicit the links between fossil fuel consumption and the retention of additional heat through the greenhouse effect. But more than simply showing that the CO_2 emissions from coal, gasoline, and natural gas were warming the planet, the transdisciplinary nature of ESS revealed the cascade of effects playing out downstream of the warming.

By the early twenty-first century, the magnitude and totality of changes humans were imposing on the Earth system were coming into focus. Through the lens of ESS, scientists could see that we weren't just changing the planet's temperature; we were changing its evolutionary state. In 2000, this recognition was given a name when geologists Paul Crutzen and Eugene Stoermer proposed that the Earth was entering a new geologic epoch they called the Anthropocene.[21] The current epoch, called the Holocene, had begun at the end of the last ice age approximately twelve thousand years ago. What we are calling the "project of civilization"—meaning all that followed from the Neolithic agricultural revolutions occurring in human societies across the planet—happened during the relatively temperate and moist conditions of the Holocene. What Crutzen and Stoermer proposed was that the scale of human activity had reached the point where it now dominated the processes that set the state of the coupled planetary systems.

The definition of the Anthropocene was originally restricted to stratigraphy and the question of whether a clear imprint of our epoch would be imposed on Earth's rock record. Within an ESS context, however, it soon expanded into a set of questions about the boundaries for the human civilization and the Earth system. These boundaries represented limits past which our project of civilization would push Earth's multisystems behavior into an entirely different state, one very likely to be detrimental to that project. The ESS boundaries perspective made the magnitude of our impact on the planet explicit. For example, human beings now move more of the key elements nitrogen and phosphorus around the planet than all other natural processes combined.[22] We've "colonized" more than 50 percent of the planet's open land for our uses. We shifted the oceans' global pH levels, making them more acidic. We're driving a massive reduction in the biosphere's diversity that will likely constitute a sixth mass extinction event. Finally, and most obvious, we're altering the entire atmosphere's composition and radiative transfer properties through our greenhouse gas emissions.

Although the concept of the Anthropocene is now used to frame these events and has gained widespread currency outside science, the concept is not without its problems. The idea that "we live on a human dominated planet," while meant to call our attention to the widespread and devastating effects of human actions on the biosphere, is nonetheless anthropocentrically blinkered, considering that the microbes really dominate the planet, as Margulis never tired of reminding us.[23] As Bruce Clarke notes, "The concept manifestly isolates and foregrounds a species she had long been at pains to put in its place alongside its beleaguered planetmates."[24] Philosopher Kathleen Dean Moore has also called attention to the human-centered bias of the concept.[25] Another problem is that the Anthropocene narrative depicts all of humanity as a species dominating the Earth system, but the Anthropocene is not the product of all humanity, but rather of particular groups of people, in particular those responsible for Western colonialism, a key driving historical force behind the Anthropocene.[26] Kyle Whyte notes, "'Anthropogenic climate change' or 'the Anthropocene,' then, are not precise enough terms for many Indigenous peoples, because they sound like all humans are implicated in and affected by colonialism and industrialization in the same ways."[27] For these reasons, although we use the term *Anthropocene*, we also call attention to its shortcomings and to how it encodes aspects of the Blind Spot we wish to overcome.

The recognition of the Earth's entry into this new human-shaped phase, problematically labeled the Anthropocene, could not have happened without the construction of Earth systems science and its emergent transdisciplinary perspective. Earth systems science was more than just a new kind of science; it also represented a new kind of need for science. Rather than reductionism, ESS had to embrace complexity as a first principle. In addition, because human effects on the terrestrial physical system had become widespread, attempts to predict future trajectories for the system required addressing the "wicked problem" of feedback loops of perception, behavior, and consequences.

In this way, ESS revealed the nature and scope of the human transformation of the planet. It demonstrated how Earth functioned as a complex system in the past. It showed us how our activity, even from the beginning of the agrarian era, had transformed the behavior of that complex system. Most important, it revealed the span of possible trajectories flowing into the future from the present. Many of these paths yield an Earth on which it will prove difficult, if not impossible, for our project of civilization to flourish. Earth systems science showed us the existential risk we face now and laid out the actions we needed to take to mitigate that risk.

In response we did nothing.

Despite what ESS revealed to the world's scientists, the world's political and economic powers refused to take meaningful action. Though numerous treaties were signed, most notably the Paris Accords, none of the major emitter nations reduced its carbon output in significant ways. During the time we've been writing this book, fifty-five years after President Johnson warned the US Congress of global warming, the world has been experiencing multiple extreme weather events: 110 degree Fahrenheit temperatures in the Pacific Northwest, massive wildfires in California and Australia, catastrophic Atlantic hurricanes, huge floods in Northern Europe, scorching heat waves in India and Pakistan, extreme rainfall in southern Africa, and the list goes on. We are now so deep into the climate crisis that even drastic reductions in CO_2 emissions will not stave off significant warming already built into the system.

Although the reasons for this inaction are manifold and require their own detailed study, science is too powerful a force to ignore how ideas associated with it became entrenched in culture. In *Reason in a Dark Time*, philosopher Dale Jamieson discusses how politics and economics, as well as moral philosophy, failed to respond to the climate crisis.[28] Jamieson's title is telling because it highlights how a particular ideal of reason became expressed in the logic of our political and economic systems. The logical structure and the ideal of reason these systems embody weren't formed out of thin air. Instead, they were an expression of a theory of knowledge that was formulated in the seventeenth and eighteenth centuries along with what we are calling the Blind Spot. We can find the Blind Spot's ideal of reason, which claimed to speak for science, underpinning the political economy of the emerging Industrial Age. As we'll argue next, it was the culture of that mechanistic political economy and its version of the Blind Spot that drove us into the Anthropocene and that has proven unable to respond to its urgency.

The Blind Spot, Culture, and Political Economy

From the late seventeenth century to the mid-nineteenth century, the essential ideas of industrial capitalism, such as production, demand, markets, investment, risk, and profit, were being developed by thinkers such as Adam Smith and David Ricardo. At the same time, the material capacities of industry grew exponentially, driving rapid change across the European continent and around the world through European colonization. In response to the profound disruptions and disparities that followed these changes, alternatives to capitalism were proposed, most notably in the writings of Karl Marx and Friedrich Engels. Eventually these two visions of political economy, capitalism and socialism, would come to dominate the organization of the world. They were the intellectual cornerstone of a resource- and energy-intensive industrialism that was unrelenting in its transformation of human culture and, eventually, planetary dynamics.

The overlap of the rise of capitalism (and eventually socialism) with the rise of science is no accident. Science and its offspring technologies unleashed knowledge that fed directly into the stunning increase in productive capacities of the new industrial society. Newtonian physics, for example, allowed engineers to build the machines that ran the factories, while advances in chemistry allowed new generations of dyes, fermenters, and fertilizers. Although this overlap in time is obvious, the overlap of causes in the relationship between science and capitalism has not been as clear. Scholarship in the history of science and capitalism has often tended to see them as related but separate. One example of their relatedness comes from W. W. Rostrow, who saw science as helping to account for the rise of modern capitalism. According to historians Lukas Rieppel, Eugenia Lean, and William Deringer, Rostrow "identified 'the gradual evolution of modern science and the modern scientific attitude' as a decisive factor separating vibrant, capitalist economies from less dynamic predecessors and alternatives"[29] In these classic historical discussions, however, science was still portrayed as existing in a separate space as a search for pure knowledge uncorrupted by the pursuit of profits. If some scientific ideas proved useful for industrial ventures, that was seen as a peripheral outcome unrelated to the true concern of scientists. This perspective was embodied, in particular, in Robert K. Merton's famous description of the scientific community as one "whose normative structure effectively insulates its members from the demands of the marketplace."[30]

This view of the relative separation of science from the domains of political economy has been eroded by a new generation of scholars. The entangled histories of science and industrial political economies, be they capitalist or socialist, have recently emerged as a topic of studies, building off earlier concerns raised by feminist scholars, notably Donna Haraway, Sandra Harding, and Evelyn Fox Keller, and researchers within the discipline of science and technology studies (STS).[31] These writers seek to understand the intertwining of both ideas (knowledge) and materials in the coeval rise of science and a rapidly globalizing industrial culture. As Rieppel, Deringer, and Lean, the editors of the important *Osiris* volume on "Science and Capitalism: Entangled Histories," put it: "Not only should thinking, calculating, planning, forecasting, organizing, and theorizing all be afforded a central place in the history of capitalism, we contend, but these seemingly abstract and disembodied activities can and ought to be studied as genuine forms of practice with the power to produce far-reaching effects in surprisingly distant parts of the world."[32] The Earth's entry into the Anthropocene would seem to justify this claim.

In this new perspective, the emergent capitalism of the seventeenth and eighteenth centuries is seen as an epistemic system that cannot be pulled apart from those being assembled and valorized in science. As historian Harold Cook puts it, "We can say that economies and sciences go together like body and mind."[33] Speaking directly of the scientific revolution in Europe, Cooks adds, "Where merchants held power, so did the kinds of commensurable material knowledge they most valued."[34]

We thus begin with the recognition of these entangled histories of science and industrial political economy. Given the tight braiding of science and early industrial capitalism, we should expect that the former's Blind Spot, with its limited perspectives on nature, matter, life, and the value of experience, would get woven into the latter's forms of thinking, calculating, planning, and value systems. In addition, we suggest that these tightly woven threads of the Blind Spot in the political economies of industrial society are an important cause of our society's inability to comprehend the rise of the Anthropocene, its existential threats, and its profound inequities across the Global North and South, as well as between settlers and indigenous peoples.[35]

We acknowledge that this proposed codependency of science and the value system behind industrialization would require an entirely separate work to elaborate using the scholarly tools of history, economics, and political science. We offer here just a few points to motivate our claims. To see how the scientific Blind Spot became woven into the philosophical roots of industrial political economy, we return to our list of the Blind Spot's characteristic ideas and look for examples of how they also manifested in the political economy and culture of the industrial worldview.

We start with the reification of mathematical entities. Although fully mathematized theories of economics are now a standard tool in political decision making, the use of calculation in economic and political policy was not central to internal debates among European nations until after Newton. William Deringer writes in *Calculated Values: Finance, Politics, and the Quantitative Age*, "In the English-speaking world, the notion that numerical calculation was deserving of special esteem as a way of thinking and knowing, particularly in political contexts, first began to take hold in Britain around the turn of the eighteenth century. Prior to that point numerical thinking had held a rather marginal place in political affairs."[36] It was during this period that a new ideal of political reason emerged. Debates couched in numerical, and ultimately mathematical, frameworks entered the political consciousness. Before this time, however, the link between reason and mathematics was mostly the concern of philosophers, and in the context of mathematical applications to natural phenomena, natural philosophers. Deringer writes, "Up through most of the seventeenth century, quantitative thinking held no special prominence in political practice or public culture."[37]

But in the wake of the successes of science, political arguments over economic decisions began to be cast in quantitative, calculative language. William Pulteney, a member of Parliament and a political commentator, wrote in 1727, "*Facts and Figures* are the most stubborn Evidences."[38] Economic propagandist (and accused pirate) John Crookshanks made the link more explicit a few years earlier when he wrote, "Truth and Numbers are always the same."[39] By the middle of the eighteenth century, a new authority had been granted to quantitative reasoning. According to Deringer, it "marked a decisive transformation in Briton's collective *civic epistemology*."[40] The rise of quantitative reasoning during this period was not simply a matter of accounting. Within political economy, it was the beginning of a mathematization of both ontology and epistemology. What was known, and how to know it, depended on the ability to cast it into mathematical terms, to the point that mathematics would soon take on an ontological life of its own.

Thus, a quantitative mathematical view of political economy appeared alongside the quantitative mathematical view of nature that was becoming triumphant during this period. The effect downstream to our own era is an economics that is fully mathematical. Practitioners of modern economic theory are expected to be adroit in integral calculus and deploy its tools of mathematical reasoning and modeling. This use of mathematical reasoning need not, by itself, be a problem. When, however, the mathematical variables in economic theory are taken to represent independent "things," rather than seen as reductive abstractions of experienced reality, and the transitional act of reductive abstraction is forgotten, we have another case of surreptitious substitution, the fallacy of misplaced concreteness, and the amnesia of experience. We end up with a powerful tool that is dangerously limited.

To understand the danger, and its connection to Blind Spot metaphysics, consider so-called neoclassical economics, the economist's version of the physicist's Standard Model. Two essential components of neoclassical economics are the idea of rational actors and the efficient-markets hypothesis (EMH). Individuals are considered to behave rationally in their own economic self-interest, and market prices are thought to fully reflect all available information. Neoclassical economic theory holds that individual rational agents, when aggregated, act with perfect information and efficiently allocate capital. In this way, markets can price things accurately. The achievement of this efficient allocation happens when equilibria between various utility functions occur (supply and demand, for example). Such equilibria are mathematically expressed as the intersections of the functions at particular price points.

EMH, along with its assumption of rational actors, has been strongly criticized for regarding human behavior (and experience) in highly abstract and idealized ways that barely resemble reality. These criticisms matter for our argument because they point to how neoclassical economics embodies

Blind Spot reductionism, objectivism, and the reification of mathematical constructs. The theory posits highly simplified "economic atoms," from which mathematical rules can be derived about the behavior of the system as a whole. All rational actors are identical in that their desires to optimize their economic gain, and hence their behaviors, are the same. EMH reifies the utility functions, attaching to them a reality in economic debates that excludes real-world characteristics that cannot be captured in such functions. This conception is most apparent in the long-standing debate about how to deal with "externalities" in neoclassical economics. The biosphere and its functions are the most important externalities for our concern. The inability of neoclassical models to even recognize the biosphere's existence can be seen as their most significant Blind Spot and is a strong motivating reason for alternative models, such as environmental economics and ecological economics.

When it comes to dealing with climate change, this limitation of the neoclassical view has become all too apparent. William Nordhaus's work, for which he was awarded a Nobel Prize in economics in 2018, is often considered the best application of the neoclassical view to climate change. This application, however, left Nordhaus accepting a 3.5 degrees Celsius increase in global temperatures as being optimal for the economy.[41] This result was based on modeling, from a neoclassical perspective, how different sectors of the economy would respond to increases in temperatures. We have begun to see, however, how just the 1.1 degrees of warming that has already occurred has affected not just the economy but also the social order. These experienced realities have made Nordhaus's calculations seem not only fatally wrong but also Dickensian in assuming the economy will keep on churning along no matter how much social disruption climate change causes.

Embodying the Blind Spot's materialism and bifurcation of nature, neoclassical economics and the political economy built from it do not know how to place value on the world except as physical resources for production (once again, this is the problem of externalities). This failure has led to a long-running debate among those who advocate for action on climate change about how to consider the issue of nature's value. Recognizing the power wielded by neoclassical economic views, some people suggest that the problem of climate change must be posed in terms of the "ecosystem services" the biosphere provides the economy, the idea being that only by pricing these services correctly with standard economic models will the political economy finally be able to grasp the value of the biosphere. Although there are some good points to be made in this perspective, it deprives political economy of the understanding that the human economy is subservient to the biosphere, not the other way around.

The failure of the political economy that drove the industrial age's great acceleration is all too familiar. Dale Jamieson notes that the problem with our economics when it comes to climate change "is not that it fails to have the right numbers but that there is more at stake than what the numbers reveal."[42] But numbers can't reveal entities that exist outside the reductionist, materialist ontology of our economics. Jamieson tells us that "economics alone cannot tell us what to do in the face of climate change."[43] But in a Blind Spot worldview where everything reduces to "nothing but" objectified physical entities, whether they be fundamental particles or abstract rational actors, we are left without an account of what really matters. Jamieson continues, "Not all of the [economic] calculations can be performed, and even if they could, they would not tell us everything we need to know."[44] If there are important aspects of our world—the experienced world—that do not fit into the physicalist picture of reality, including its rendering in economics, then there will be aspects of that which we must value to which we will remain insensitive. That is why the political economies of the industrial age have driven society to such rapacious consumption and why they have been entirely unable to respond to the dangers that consumption poses even as these dangers stare us in the face. In the end, Jamieson says, "At its best economics is a science and therefore cannot tell us what to do. At its worst it is an ideology, a normative outlook disguising itself as a report on the nature of things."[45] Our concern has been to call attention to how the Blind Spot has both shaped and been shaped by that ideology.

Complex Systems Science

Science's ability to reveal deep patterns in the natural world and provide the means to better the human condition is rightly celebrated. But science and the technologies it has yielded have also played a pivotal role in creating the plight of the Anthropocene. This happened in a direct way by providing the means to drive industrial-scale resource extraction and consumption. It also happened, as we have just argued, in an indirect way via the cultural, political, and economic adoption of a worldview that was mistakenly identified with science, the worldview we call the Blind Spot.

Science, however, is not the same as this worldview. The Blind Spot metaphysics and theory of knowledge may have come to be attached to much of science in our culture, but science is separable from them. Science strives to be a self-correcting narrative built from a communal, open-ended, inquiry-focused dialogue with experience. It cannot be bound by the philosophical preferences of its practitioners. As the scientific workshop provides new tools for enlarging experience, science responds by opening up new pathways for its own narratives. Thus, the map is not only extended but also redrawn with new keys and symbols to represent newfound aspects of nature.

In this way, new developments in science itself have challenged the Blind Spot as a taken-for-granted conception of science and reality. We've already explored how these challenges have arisen in cosmology, quantum physics, biology, cognitive science, and the neuroscience of consciousness. We end this chapter by describing how these challenges continue with the new tools and perspectives we identified as crucial to the success of Earth systems science. In following the history of that field from Vernadsky through Lovelock and Margulis, we noted the key role of systems thinking. The rise of such a nonreductionist way of conceptualizing the world, however, was hardly confined to the study of the Earth. Instead, it represented a profound shift in the nature of scientific inquiry that continues to this day and falls under the broad title of "complex systems theory."

"The science of complex systems is not an offspring of physics, biology, or the social sciences, but a unique mix of all three."[46] So opens a textbook on complex systems theory by Stefan Thurner, Rudolf Hanel, and Peter Klimek. As we sketch the main features of the field and how it fosters perspectives that differ from those of the Blind Spot, the first point to note is its transdisciplinary character. Complex systems theory is born from a diversity of perspectives on a diversity of subjects that previously eluded science.

Complex systems can be described as coevolving multilayer networks. Unpacking this description will show how the perspectives embodied in complex systems theory differ from those of classical physics (from which the Blind Spot metaphysics was born).

We begin with coevolution. In the models of classical physics, the evolution of matter and energy are controlled by differential equations. Boundary conditions are required to solve these equations. Specifying the boundary conditions requires that we separate the system (for example, the collection of particles) from its environment. The properties of the environment are held fixed, and they are used to specify the needed boundary conditions. These are then fed into the differential equations, which, if they can be solved, determine the evolution of the system. Although it may be possible to specify a time-dependence of the boundary conditions beforehand, the system and the environment cannot evolve together. This means it's usually not possible to allow for interactions by which the system and the environment mutually and simultaneously determine each other's future states. In other words, the classical approach to physics (again, the approach from which the Blind Spot took its form), does not allow for coevolution.

Thus, biological evolution, in the Darwinian sense in which organism and environment coevolve, and thereby generate novelty, falls outside the domain of classical physics, as we argued earlier. Life is based on physics but beyond physics. In physics, "complexity" often means systems made of many particles. This is the domain of statistical mechanics. Treating most problems in that domain, however, requires that particle systems have the property known as ergodicity. Given enough time (or many identical copies of the system), all possible system states will eventually be achieved. In terms of the phase-space framework discussed in previous chapters, ergodicity means that all points in the multidimensional phase space within which the system "lives" will be explored by the system. This ergodic assumption is how statistical mechanics assigns probabilities for different evolutionary trajectories. But evolution in the biological (and cultural) sense is not ergodic, as Stuart Kauffman has long argued.[47] Instead, each evolutionary step depends on the entire history of the system and its environment, as well as the restricted possibilities available conditioned on the present. Kauffman calls this aspect of evolution "the adjacent possible."[48] The adjacent possible is "the subset of all possible worlds that are reachable within the next step and [that] depends strongly on the state of the world."[49] For systems subject to evolutionary processes, the phase space cannot be specified beforehand. The system's trajectory is path dependent (its present state depends on the path it took to get there) and depends on the adjacent possible (where it can go is conditioned on where it is). There is no preexisting God's-eye view that determines all that is possible. Life has no prestateable phase space.

As a result, we now have a way of regarding nature as fundamentally creative. Nature allows for, indeed generates, the emergence of novelty, as the coevolution of systems and environments continually opens new futures in the adjacent possible. Whitehead had already discerned this conception when he wrote, "Nature is never complete. It is always passing beyond itself. This is the creative advance of nature."[50]

Networks are the next key feature of complex systems. A network is composed of nodes and links between them. A familiar representation of a network is the map of a subway system in which each station represents a node and the train lines running between those stations represent the links. The networks in complex systems, however, are far more dynamic than subways. The links between the nodes can be broken and reformed on timescales that are short compared with the lifetime of the system. The strength, direction, and nature of the links can also change in time. Social networks are a good example of this kind of behavior. Social relations between friends and groups of friends are constantly shifting, growing stronger or becoming weaker in time. Much of the world can be described from a network perspective, including protein-protein interactions in cells, large-scale distributed activity in the brain, food webs in an ecosystem, economic relations between companies, and political relations between nation states.

The network perspective is not reductionist. It dispenses with smallism, the idea that small things and their properties are more fundamental than the large things they constitute. It also rejects microreduction, the idea that the best way to understand a complex system is to break it up into its elements and explain the properties of the whole in terms of the properties of the parts. The nodes of a complex system are not considered to be atoms whose prior specified properties determine the network's architecture. Instead, the network's architecture and behavior are emergent properties that arise from the collective activity of the entire system, its environment, and their coevolving history. Knowing only the properties of the nodes before the dynamics begins will not deliver the network's evolutionary trajectory. One reason is that the interactions between the nodes can evolve with the network. In contrast, Newton's prescription for force or gravity between a collection of stars does not evolve in time. But in a network of organisms in an ecosystem, the nature and the strength of the interaction will change depending on how the network evolved into its current state as well as that state's configuration.

Equally important, most complex systems involve multiple layers of networks. In considering responses to climate change, for example, researchers must include the biogeophysical network comprising the Earth system, the energy systems that power human civilization, the transportation networks that supply those energy systems, the economic networks that fund the transportation systems, and the social/political networks that respond to changes in the biogeophysical networks. Thus, each node in each network layer may appear again in other network layers, with connections within and between the layers. Because key aspects of the total dynamics emerge only from the large-scale interactions, it is often impossible to "see" the behavior possible in multilayer networks just from a consideration of node properties. It is often impossible even to specify the node properties until they appear within these networks, since network interactions allow the nodes to take on certain properties. Thus, local and global elements and patterns are mutually determining, as we saw in our discussion of life.

The tools used to explore complex systems are also noteworthy. Here the key role of the scientific workshop makes another appearance. Complex networks must almost always be explored through computer simulations rather than through analytic means. Whereas gravity always operates in interaction between masses, certain nodes in a protein-protein network will "turn on" only if global conditions in the network have evolved to allow them to do so. There is no way to analytically calculate exact solutions for this kind of dynamics. Instead, the behavior must be simulated computationally. For this reason, the fundamental tool of complex systems theory is the computer.

Before digital computers, the mathematics that could be brought to bear on system models was composed, for the most part, of linear differential equations. Linear systems are ones in which there is a direct proportion between changes in the inputs and changes in the output. Most of the systems that science studies, however, are nonlinear. Indeed, mathematical physicist Stanislaw Ulam is said to have remarked, "Using a term like nonlinear science is like referring to the bulk of zoology as the study of non-elephant animals."[51] In general, nonlinear equations cannot be solved analytically (that is, by framing the problem in a well-defined form and finding a unique solution), except in special cases. With the advent of digital computers, however, nonlinear systems could be explored by simulation (that is, by using step-by-step methods to explore the approximate behavior of the mathematical model of a system on a computer). Using this approach, problems such as those in fluid mechanics that had remained intractable for centuries were finally opened to scientific investigation. It's noteworthy that simulations of fluid dynamics were essential to understanding Earth's climate via the advent of general circulation models in the 1980s.

Complex systems composed of multilayer networks are inherently nonlinear. The nature of their interactions is such that in many cases, it may not even be possible to write down initial sets of differential equations to represent them. For these reasons, complex systems science relies on computer simulations. Thus, it was only with the advent of cheap, easily accessible tools for simulations that the richness of complex systems as a transdisciplinary science could mature. Today complex systems methods are being deployed in a wide variety of settings from understanding the geophysical evolution of planets, the origins of life, the behavior of cells, the workings of the brain, the dynamics of social systems, and the behavior of stock markets.

We have ended this chapter with a short sketch of complex systems science, particularly as it arises in the context of Earth systems science, not because we think it can take us beyond the Blind Spot all by itself, let alone lead us to a better version of human society in the face of the environmental and social challenges we face. As with classical science, the technologies emerging from complex systems science can be deployed in ways that cause great harm. The impact of social media on democratic norms is a clear example. It would also be naive and foolish to think that complex systems science is not without its own blind spots. Rather, we have highlighted it because it challenges the reductionism built into the Blind Spot. By embracing, from the outset, a nonreductionist perspective that focuses on the centrality of entangled, looping relations and their emergent properties in complex systems, with ourselves included as inextricable from the entangled loops we study, complex systems science offers a glimpse of what a scientific worldview beyond the Blind Spot might look like. We think this is a hopeful sign.

Afterword

We began this book by pointing to a paradox. Science tells us that we human beings are peripheral in the cosmic scheme of things, but also that we're central to the reality we uncover. We're a tiny contingent presence in a vast universe, yet our experience is ubiquitous in scientific knowledge. We've seen many examples in the previous pages.

Instead of trying to avoid this paradox, we can and should embrace it. We're the authors of the scientific narrative, and we're characters within it. As authors, we create science. As characters in the narrative, we're a minuscule part of the immense cosmos. This is how we must portray ourselves based on what we've discovered in cosmology and biology. Still, we must not forget that these tiny characters—we ourselves—are also the authors of the narrative. We've written it using the tools we created as a community of learners in the collective scientific workshop. Such is the strange loop of scientific knowledge.

We have argued that we must inscribe ourselves back into the scientific narrative as its creators. Science rests on how we experience the world. There is no way to take ourselves out of the story and tell it from a God's-eye perspective. Forgetting this fact means succumbing to the Blind Spot, and that means losing our way in both science and all the critical ways that science shapes society.

Our aim has been to call attention to the Blind Spot. We've done this by delineating its central ideas and charting their presence across the scientific fields that inform our worldview. We've described how the constellation of overlapping ideas we call the Bind Spot arose over the course of many centuries and increasingly shaped how the scientific endeavor was understood. But as we have argued, the scientific project and the Blind Spot worldview are not the same thing, and the advance of scientific knowledge in the twentieth and twenty-first centuries demands that we leave the Blind Spot behind.

What, however, does leaving the Blind Spot behind look like? Although we've pointed to many scientific ideas that go beyond the Blind Spot, we have not tried to formulate a comprehensive scientific or philosophical perspective to replace it. This choice is deliberate. Science is a collective project. It is the entangled dance of collaborative theory building and experimentation that ultimately moves science forward, not critical reflection in one's study. Our hope is that by bringing the Blind Spot into our collective vision, we will be better able to find new paths beyond it, so that our scientific civilization can survive and flourish in this millennium. Still, in closing we would like to suggest what best practices might look like now, for researchers and the broader public, as together we strive to explore where and how new efficacious perspectives for our future may emerge.

The first step along a path beyond the Blind Spot is a historical and philosophical awareness of how present theory building in science carries the weight of its past perspectives. As research communities progress, they often face junctures when decisions between various explanatory frameworks are made. The split between orthodox versus heterodox viewpoints will often rest on assumptions that remain unexamined. Sometimes these assumptions are purely scientific and methodological. At other times, however, they rest on philosophical commitments that have not been brought to the surface. These commitments, which lie submerged, get handed down from professor to student across generations until they simply become a field's orthodoxy, "what everybody knows." By bringing this history to light, these barely articulated philosophical commitments can be seen as such. They can then be debated as a collection of ideas that affect scientific practice while not always being a necessary part of it. In the end, the community may choose to retain those commitments or begin to explore alternatives.

We have spent most of this book trying to carry out the first part of the project outlined above—excavating and laying out history and submerged philosophy to understand how Blind Spot metaphysics came to permeate so much of the background in which modern scientific practice is conducted. We have done so in the hope that this will open gateways for the second part of the best-practices process—exploring alternatives. Thus, along the way, we have examined various ideas and trends in modern science that seem to represent possible gateways into post–Blind Spot perspectives: network theory, complex systems theory, QBism, relational quantum mechanics, biological autonomy, embodied cognition, enaction, and neurophenomenology. We are certainly not claiming that they represent a coherent philosophical and methodological alternative to the Blind Spot. Such an alternative has yet to be fully worked out. That effort, we would argue, is, or should be, part of the principal project for twenty-first century science and philosophy of science. With regard to best practices now, however, there are some general recommendations worth outlining.

Recall that we have explicated four pathologies associated with the Blind Spot: (1) surreptitious substitution, (2) the fallacy of misplaced concreteness, (3) reification of structural invariants, and (4) the amnesia of experience. Thus, as researchers face choices within their own domains or attempt to evaluate various theoretical perspectives and methods, the simplest prescription is, "Don't repeat those mistakes." Be aware of the perspectives and procedures that support or reinforce these pathologies and look for alternatives that don't lead to them. These alternatives will not mistake abstract theoretical constructs for concrete lived experience. They will not forget that the ascending spiral of abstraction is always rooted in the primacy and irreducibility of experience. They will also recognize the process by which aspects of embodied experience are isolated and extracted so that they can be transformed into the structural invariants on which the abstractions are built. To put it in a nutshell, look for alternatives that can recognize and embrace the strange loops of experience and science rather than favoring approaches that ignore or try to obviate them.

Note that this doesn't mean rejecting abstractions. Theoretical structures, often represented in the language of mathematics, are one of science's most potent tools. As Whitehead argued, abstractions are not the problem. The problem is failing to understand their nature and surreptitiously substituting them for the concrete. Thus, within their own domains, researchers need to ask whether an interpretation or reading of the abstractions misconstrues human experience and the life-world. Or does the interpretation under review understand, or at least acknowledge, that the abstractions arise from, and depend for their meaning on, the life-world and the collective scientific workshop?

Implementing this best practice becomes more urgent the closer the scientific topic lies to the advancing edge of research and social policy. In a laboratory study of the electrical conductivity of minerals, for example, questions about the implications of different philosophical commitments may not be as relevant (though even here, questions can still arise). But in domains where such commitments can shape entire research agendas, such as cosmology, the foundations of quantum mechanics, and the scientific study of consciousness, or can have huge social consequences, such as AI, interrogating the underlying philosophical commitments of the research becomes imperative.

So far in this brief exploration of best practices, we have focused on the work within science and the philosophy of science. Of equal importance are consideration of best practices in how science is represented to society at large. Here the Blind Spot has great impact. There are many ways in which the philosophical commitments of the Blind Spot are sold to the public as "what science says." It may take the form of science documentaries telling people they are nothing more than their so-called genetic programming (genes aren't programs, and they require the existence of whole organisms embedded in their ecosystems to be expressed). It may be breathless science news articles that claim future generations will upload themselves into computers (your selfhood or personhood isn't a computational data structure). It may be public lectures or op-eds that claim physics has now answered the question of why there is something rather than nothing (this is not the kind of question science can answer). Such stories express a form of scientific triumphalism based on the Blind Spot. Those who argue from this perspective may claim that it's the only way to safeguard science from the forces that oppose it. While we agree that safeguarding science is essential, especially in our age of rampant science denial and intentional disinformation, doubling down on Blind Spot commitments does not serve this end and may well backfire.

Indeed, the scientific triumphalism based on the Blind Spot is not a harmless exaggeration but a harmful overreach. It feeds the stereotype of the scientist as cold, emotionless, and "not like us" (as one recent poll expressed it). When Blind Spot ideas are presented to the public as facts that only the naive and uneducated would dispute, it is likely to exacerbate opposition to science in public policy debates (for example, about climate change mitigation or vaccination). Furthermore, as we have argued, representing nature or the environment as essentially a resource is tied to the elevation of objective quantitative measures as more fundamental than the human experience that lies at their source. The long history of inaction on climate change can be traced in part to the inability to change this mindset in which the Earth becomes a set of "ecosystem services" that can be monetized as value provided to the technosphere.

Thus, best practices in the domain of science and society include becoming aware of how the story of science is told to the public. Without doubt that story is about the profound capacity of the human imagination and our ability to prevail over ignorance and bias. But if the story is told as one of transcending the human, then it becomes an essentially religious narrative about the search for a perfect knowledge beyond our finitude. Instead of saying that science is a means for rising above the great, strange mystery of being human in the vast wide world, a better story is that science takes us deeper into that mystery, revealing new ways to experience it, delight in it, and, most of all, value it. By leaving behind the Blind Spot, we can properly understand the crucial importance of objectivity as a means for public knowledge without transforming it into a dubious ontology. Most important, we can appreciate just how remarkable the human activity of science is and how necessary is the fight for its integrity, without making it a proxy for centuries-old philosophical beliefs that are no longer relevant to where are and where we need to go.

Acknowledgments

This book was born from conversations among the three of us that began when Marcelo Geiser invited Adam Frank and Evan Thompson to be visiting fellows at the Institute for Cross-Disciplinary Engagement (ICE) at Dartmouth College in 2017. A subsequent workshop, "The Blind Spot: Experience, Science, and the Search for Truth," took place at ICE in 2019. We thank Dartmouth College and the Templeton Foundation for supporting ICE and making these meetings possible. We also thank the workshop participants—Michel Bitbol, Chris Fuchs, Jenann Ismael, Peter Lewis, Michela Massimi, Robert Sharf, Mark Sprevak, and Peter Tse—for their friendly and helpful critical engagement with our ideas.

Another helpful and stimulating testing ground was a 2022 conference, "Buddhism, Physics, and Philosophy Redux," held at the Center for Buddhist Studies at the University of California, Berkeley, and organized by Robert Sharf. The conference participants—Michel Bitbol, Craig Callender, John Dunne, Chris Fuchs, Jay Garfield, Jenann Ismael, Huw Price, Carlo Rovelli, Robert Sharf, Francesca Vidotto, and Jessica Wilson—offered many helpful critical responses to our ideas.

Several people read parts of the manuscript at different stages of the writing and gave us helpful advice: Michel Bitbol, Robert Crease, Jay Garfield, Alyssa Ney, Robert Sharf, and Rebecca Todd. Two anonymous reviewers for MIT Press offered useful suggestions for improvement. We also thank the Philosophy graduate students at the University of British Columbia who read and responded to the manuscript: Tyeson Davies Barton, Laura Bickel, Albert Cotguno, Elena Holmgren, Alexandra Jewell, Emily Lawson, Jelena Markovic, Anthony Nguyen, Matthew Perry, Ying Yao, Ceren Yildiz, Yuki Ueda, Davide Zappulli, and Angela Zhao.

An early version of the book's core idea was published as an online article in *Aeon* (2019). We are grateful to our editor, Sally Davies, for helping us
improve the writing and for her dedication to the article.

Our final thanks go to our literary agents and editor. Howard Yoon at
the Ross Yoon Agency represented the book and found an excellent home
for it in the hands of editor Philip Laughlin at the MIT Press. We also thank
Michael Carlisle (Marcelo Gleiser's agent) and Anna Ghosh (Evan Thompson's agent) for their help and support.

Notes

An Introduction to the Blind Spot

1. See Jonardon Ganeri, "Well-Ordered Science and Indian Epistemic Cultures," *Isis* 104 (2013): 348–359.

2. Maurice Merleau-Ponty, *Phenomenology of Perception*, trans. D. Landes (London: Routledge Press, 2013), 84.

3. See Hasok Chang, *Inventing Temperature: Measurement and Scientific Progress* (Oxford: Oxford University Press, 2007).

4. Robert Crease, *The Workshop and the World: What Ten Thinkers Can Teach Us about Science and Authority* (New York: Norton, 2019).

5. Thomas Nagel, *The View from Nowhere* (New York: Oxford University Press, 1986).

6. See Jimena Canales, *The Physicist and the Philosopher: Einstein, Bergson, and the Debate That Changed Our Understanding of Time* (Princeton, NJ: Princeton University Press, 2015).

7. Michel Bitbol also argues that the Blind Spot of our scientific worldview is pure experience. See Michel Bitbol, "Beyond Panpsychism: The Radicality of Phenomenology," in *Self, Culture, and Consciousness*, ed. S. Menon, N. Nagaraj, and V. Binoy (Singapore: Springer, 2017), 337–356.

8. William James, "The Thing and Its Relations," *Journal of Philosophy, Psychology and Scientific Methods* 2, no. 2 (1905): 29.

9. Kitarō Nishida, *An Inquiry into the Good*, trans. Masao Abe and Christopher Ives (New Haven: Yale University Press, 1992).

10. See Fujita Masakatsu, "The Development of Nishida Kitarō's Philosophy: Pure Experience, Place, Action-Intuition," in *The Oxford Handbook of Japanese Philosophy*, ed. Bret W. Davis (Oxford: Oxford University Press, 2019), 389–415.

11. Nishida, *An Inquiry into the Good*, 7.

12. When our book was already written we discovered William Byers, *The Blind Spot: Science and the Crisis of Uncertainty* (Princeton, NJ: Princeton University Press, 2011). Byers's idea of the blind spot and his analysis of science converge remarkably with ours.

Chapter 1

1. Edmund Husserl, *The Crisis of European Sciences and Transcendental Phenomenology*, trans. David Carr (Evanston, IL: Northwestern University Press, 1970), 14.

2. Galileo, "The Assayer," in *Discoveries and Opinions of Galileo*, trans. Stillman Drake (New York: Anchor Books, 1957), 237–238.

3. Husserl, *The Crisis of European Sciences*, 51. Italics in original.

4. See Lee Hardy, *Nature's Suit: Husserl's Phenomenological Philosophy of the Physical Sciences* (Athens: Ohio University Press, 2013).

5. Nancy Cartwright, *How the Laws of Physics Lie* (Oxford: Clarendon Press, 1983).

6. Nancy Cartwright, *The Dappled World: A Study of the Boundaries of Science* (New York: Cambridge University Press, 1999).

7. Robert Crease, *The Workshop and the World: What Ten Thinkers Can Teach Us about Science and Authority* (New York: Norton, 2019).

8. See Crease, *The Workshop and the World*, chap. 1.

9. Carolyn Merchant, *The Death of Nature: Women, Ecology, and the Scientific Revolution* (New York: Harper, 1980), 80, 164.

10. Carolyn Merchant, "The Scientific Revolution and *The Death of Nature*," *Isis* 97 (2006): 515.

11. Crease, *The Workshop and the World*, 211.

12. Cartwright, *The Dappled World*, 2–3.

13. Cartwright, 25.

14. Crease, *The Workshop and the World*, 212–213. See also William Finnegan, *Barbarian Days: A Surfing Life* (New York: Penguin, 2015).

15. Michel Bitbol, "Is Consciousness Primary?," *NeuroQuantology* 6 (2008): 53–72.

16. John Locke, *An Essay Concerning Human Understanding*, ed. Roger Woodhouse (London: Penguin, 1997), bk. 2, chap. 8, sec. 21, 137–138.

17. See Hardy, *Nature's Suit*.

18. Cartwright, *How the Laws of Physics Lie*, essay 4.

19. For Husserl's discussion of indicative signs, see Edmund Husserl, *Logical Investigations*, trans. J. N. Findlay (London: Routledge Press, 1970), Investigation I: Expression and Meaning, 181–203.

20. Ian Hacking, *Representing and Intervening: Introductory Topics in the Philosophy of Natural Science* (Cambridge: Cambridge University Press, 1983), 22–23.

21. Cartwright, *The Dappled World*, 34.

22. Alfred North Whitehead, *The Concept of Nature* (Cambridge: Cambridge University Press, 1920; Ann Arbor: University of Michigan Press, 1957). Citations refer to the 1957 edition.

23. Whitehead, *The Concept of Nature*, 30–31.

24. Arthur Eddington, *The Nature of the Physical World* (Cambridge: Cambridge University Press, 1928), xi–xii.

25. Whitehead, *Concept of Nature*, 29.

26. Whitehead, 3.

27. Whitehead, 29.

28. Whitehead, 40.

29. Whitehead, 30.

30. Isabelle Stengers, *Thinking with Whitehead: A Free and Wild Creation of Concepts*, trans. Michael Chase (Cambridge, MA: Harvard University Press, 2011), 76, 136.

31. Alfred North Whitehead, *Science and the Modern World* (New York: Macmillan, 1925), 51, 55, 58.

32. Whitehead, 51.

33. Whitehead, *Concept of Nature*, 41.

34. Stengers, *Thinking with Whitehead*, 34.

35. Whitehead, *Concept of Nature*, 44. See also 32: "The reason why the bifurcation of nature is always creeping back into scientific philosophy is the extreme difficulty of exhibiting the perceived redness and warmth of fire in one system of relations with the agitated molecules of carbon and oxygen, with the radiant energy from them, and with the various functionings of the material body. Unless we produce the all-embracing relations, we are faced with a bifurcated nature: namely, warmth and redness on one side, and molecules, electrons and ether on the other side."

36. Whitehead, 44.

Chapter 2

1. We rely on E. A. Burtt, *The Metaphysical Foundations of Modern Science* (Mineola, NY: Dover, 2003); R. G. Collingwood, *The Idea of Nature* (Clarendon: Oxford University Press, 1945); E. J. Dijksterhuis, *The Mechanization of the World Picture*, trans. C. Dikshoorn (Clarendon: Oxford University Press, 1961).

2. Alfred North Whitehead, *Science and the Modern World* (New York: Macmillan, 1925), 7.

3. See Edward Grant, *The Foundations of Modern Science in the Middle Ages: Their Religious, Institutional, and Intellectual Contexts* (Cambridge: Cambridge University Press, 1996). See also Peter Harrison, *The Fall of Man and the Foundations of Science* (Cambridge: Cambridge University Press, 2007).

4. For translations and discussion of the pre-Socratic philosophers, see G. S. Kirk, J. E. Raven, and M. Schofield, *The Presocratic Philosophers: A Critical History with a Selection of Texts*, 2nd ed. (Cambridge: Cambridge University Press, 1983).

5. See Collingwood, *Idea of Nature*, 29–30.

6. For discussion of Empedocles, including a translation of "On Nature," see Kirk et al., *Presocratic Philosophers*, 280–332.

7. Aristotle, *Metaphysics*, trans. C. D. C. Reeve (Indianapolis: Hackett, 2016), 8, 10.

8. See Dijksterhuis, *Mechanization of the World Picture*, 22–24.

9. Dijksterhuis, 82.

10. See Dijksterhuis, 8–13.

11. David Lindberg, *The Beginnings of Western Science: The European Scientific Tradition in Philosophical, Religious, and Institutional Context, Prehistory to A.D. 1450*, 2nd ed. (Chicago: University of Chicago Press, 2007), 77.

12. A description of how Epicurian ideas of the swerve made their way into late medieval Europe can be found in Stephen Greenblatt's *The Swerve: How the World Became Modern* (New York: Norton, 2011).

13. Aristotle, *Metaphysics*, 11.

14. Collingwood, *Idea of Nature*, 50.

15. Collingwood, 55–56.

16. Collingwood, 55–72.

17. See Francis M. Cornford, *Plato's Cosmology* (Indianapolis: Hackett, 1997); Plato, *Timaeus*, trans. Donald J. Zeyl (Indianapolis: Hackett, 2000).

18. Collingwood, *Idea of Nature*, 72–79.

19. Lindberg, *Beginnings of Western Science*, 37.

20. David Lindberg, *The Beginnings of Western Science*, 1st ed. (Chicago: University of Chicago Press, 1992), 282. This sentence does not appear in the second edition of this book.

21. See Lindberg, *Beginnings of Western Science*, 2nd ed., 299–300.

22. Lindberg, 2nd ed., 308.

23. Johannaes Buridanus, *Questiones in Metaphysicam Aristotelis*, bk. XII, question 9. Edition of Iodocus Badius Ascensius, Paris, 1518; fol. 73 recto. For an account of Buridan's discovery of the law of inertia, see E. A. Moody, "Laws of Motion in Medieval Physics," *Scientific Monthly* 72 (1951): 18–23.

24. Whitehead, *Science and the Modern World*, 9.

25. Robert Crease, *The Workshop and the World: What Ten Thinkers Can Teach Us about Science and Authority* (New York: Norton, 2019).

26. For discussion of what Bacon may have meant by "vexing" nature, see Carolyn Merchant, "Francis Bacon and the 'Vexations of Art': Experimentation as Intervention," *British Journal for the History of Science* 46, no. 4 (2013): 551–559.

27. See Eric Schliesser, "Why Does Newton Use 'Axiom or Laws of Motion,'" *Digressions & Impressions* (blog), January 26, 2016, https://digressionsnimpressions .typepad.com/digressionsimpressions/2016/01/why-does-newton-use-axioms-or-laws -of-motion.html.

28. Einstein made these remarks in a conversation with Esther Salaman, a student of physics. See Max Jammer, *Einstein and Religion: Physics and Theology* (Princeton, NJ: Princeton University Press, 1999), 123.

29. See Geoffrey Gorham, "Newton on God's Relation to Space and Time: The Cartesian Framework," *Archiv für Geschichte der Philosophie* 93 (2011): 281–320.

30. As quoted in Gorham, "Newton on God's Relation to Space and Time," 312.

31. We borrow the terms "the view from nowhere" from Thomas Nagel, *The View from Nowhere* (New York: Oxford University Press, 1986), and "the absolute conception of the world" from Bernard Williams, *Descartes: The Project of Pure Inquiry* (London: Pelican, 1978).

32. Quoted in Jennifer Coopersmith, *The Lazy Universe: An Introduction to the Principle of Least Action* (Oxford: Oxford University Press, 2017), 25.

33. See Alfred North Whitehead, *The Concept of Nature* (Cambridge: Cambridge University Press, 1920; Ann Arbor: University of Michigan Press, 1957), and Whitehead, *Science and the Modern World*.

34. See Eugene Wigner, "The Unreasonable Effectiveness of Mathematics in the Natural Sciences," *Communications on Pure and Applied Mathematics* 13, no. 1 (1960): 1–14.

Chapter 3

1. Aristotle, *Physics*, bk. IV, chaps. 10–14. For a translation, see Joe Sachs, *Aristotle's Physics: A Guided Study* (New Brunswick, NJ: Rutgers University Press, 1998).

2. Henri Bergson, *Time and Free Will: An Essay on the Immediate Data of Consciousness*, trans. F. L. Pogson (Mineola, NY: Dover, 1913).

3. See Alfred North Whitehead, *The Concept of Nature* (Cambridge: Cambridge University Press, 1920; Ann Arbor: University of Michigan Press, 1957), chap. 3. Citations refer to the 1957 edition.

4. Isabelle Stengers, *Thinking with Whitehead: A Free and Wild Creation of Concepts*, trans. Michael Chase (Cambridge, MA: Harvard University Press, 2011), 56.

5. Bergson, *Time and Free Will*, 106, 107.

6. Sachs, *Aristotle's Physics*, 122.

7. Bergson, *Time and Free Will*, 107–110.

8. Edmund Husserl, *On the Phenomenology of the Consciousness of Inner Time (1893–1917)*, trans. John Barnett Brough (Dordrecht: Springer, 1991), 32.

9. For an accessible reference on the FOCS-1 see https://cmte.ieee.org/future directions/2018/12/04/n-amazingly-accurate-atomic-clock/.

10. William James, *The Principles of Psychology* (Cambridge, MA: Harvard University Press, 1981), 573–574.

11. See Marc Wittmann, "Moments in Time," *Frontiers in Integrative Neuroscience* 5 (2011): 66, https://doi.org/10.3389/fnint.2011.00066; Marc Wittmann, "The Inner Sense of Time: How the Brain Creates a Representation of Duration," *Nature Reviews Neuroscience* 14 (2013): 217–223.

12. See Marc Wittman, *Felt Time: The Psychology of How We Perceive Time*, trans. Erik Butler (Cambridge, MA: MIT Press, 2016).

13. Whitehead, *Concept of Nature*, 53–54.

14. For Aristotle's complicated views on the heavens and time, see Ursula Coope, *Time for Aristotle* (New York: Oxford University Press, 2005).

15. These are examples of laws in time, kinematic and dynamic. Such laws must be contrasted with conservation laws—those that reflect the conservation (in time) of a quantity, such as the conservation of energy or electric charge in subatomic

processes. The laws of physics reflect in their very nature the tension between becoming and being.

16. Implicit in Newton's assumption of a single universal time is that information travels at infinite speeds. This was at the core of his action-at-a-distance notion that gravity acted instantaneously across the vastness of space. We now know that this is not the case, given that the speed of light is finite and gravitational perturbations travel at the speed of light. But in Newton's time, it made a lot of sense as an approximation.

17. Isaac Newton, Scholium to the Definitions in *Philosophiae Naturalis Principia Mathematica*, bk. 1 (1689), trans. Andrew Motte (1729), rev. Florian Cajori (Berkeley: University of California Press, 1934), 6.

18. Alfred North Whitehead, *Science and the Modern World* (New York: Macmillan, 1925), 44.

19. Simple Newtonian models of planetary orbits do not need to include the hot and viscous details of planetary formation; also, they tend to neglect gravitational tidal forces unless used over long timescales.

20. When shapes are relevant, objects are treated as solid bodies, with properties obtained from summing (integrating, for continuous bodies) over many particles.

21. George F. R. Ellis, "Physics in the Real Universe: Time and Spacetime," *General Relativity and Gravitation* 38 (2006): 1797–1824.

22. The physicist Nicolas Gisin and collaborators have explored this point in several recent papers, calling it the "principle of infinite precision." See Flavio Del Santo and Nicolas Gisin, "Physics without Determinism: Alternative Interpretations of Classical Physics," *Physical Review* A 100 (2019): 062107, https://doi.org/10.1103/PhysRevA.100.062107. See also George F. R. Ellis, Krysztof A. Meissner, and Hermann Nicolai, "The Physics of Infinity," *Nature Physics* 14 (2018): 770–772, https://doi.org/10.1038/s41567-018-0238-1.

23. David Hilbert, *David Hilbert's Lectures on the Foundations of Arithmetics and Logic, 1917–1933*, ed. W. Ewald and W. Sieg (Heidelberg: Springer, 2013), 730.

24. For a brief history of heat and thermodynamics, see Marcelo Gleiser, *The Dancing Universe: From Creation Myths to the Big Bang* (Lebanon, NH: Dartmouth College Press, 2005), chap. 6.

25. This accounting doesn't include "internal" variables, such as the possible rotational motions of the molecule. If there are m such rotational motions, the degrees of freedom increase to $6 + m$.

26. This equilibrium average velocity is given by the Maxwell-Boltzmann distribution and is proportional to the square root of the temperature T.

27. Hans Reichenbach, *The Direction of Time* (New York: Dover, 1956), 113.

28. A. Connes and C. Rovelli, "Von Neumann Algebra Automorphisms and Time-Thermodynamics Relation in Generally Covariant Quantum Theories," *Classical Quantum Gravity* 11 (1994): 2899–2917.

29. Carlo Rovelli, *The Order of Time* (New York: Riverhead Books, 2018), 34.

Chapter 4

1. See Alfred North Whitehead, *Science and the Modern World* (New York: Macmillan, 1925), chaps. 3–6.

2. Locke gives different lists of the primary qualities, so it's not clear exactly which qualities he thinks are primary. See Robert A. Wilson, "Locke's Primary Qualities," *Journal of the History of Philosophy* 40 (2002): 201–228.

3. Whitehead, *Science and the Modern World*, 54.

4. Whitehead, vii.

5. Whitehead, 18.

6. Alfred North Whitehead, *Process and Reality*, corrected ed. David Ray Griffin and Donald W. Sherburne (New York: Free Press, 1978), 20.

7. A circulating electric charge creates a magnetic field perpendicular to its direction of motion, like an arrow going through the middle of a target.

8. David Z. Albert, *Quantum Mechanics and Experience* (Cambridge, MA: Harvard University Press, 1994).

9. For example, if we have an initial superposition where both states contribute equally, the values for the parameters a and b at $t = 0$ would be $a = b = \sqrt{2}/2$. To get the probabilities of finding the particle in either state requires squaring the coefficients, so that $a^2 = b^2 = \frac{1}{2}$, or 50 percent each. (For the experts, the a and b coefficients may be complex numbers. In that case, the square is actually the absolute value of the complex numbers.)

10. For the experts, EPR used the position-representation of quantum mechanics, where the state function is commonly known as the wave function, represented by the Greek letter Ψ (pronounced "psi").

11. Albert, *Quantum Mechanics and Experience*.

12. Oliver Morsch, *Quantum Bits and Quantum Secrets: How Quantum Physics Is Revolutionizing Codes and Computers* (New York: Wiley-VCH, 2008).

13. N. David Mermin, "What's Wrong with This Pillow?," *Physics Today* 42 (1989): 9–11.

14. G. C. Ghirardi, A. Rimini, and T. Weber, "Unified Dynamics for Microscopic and Macroscopic Systems," *Physical Review D* 34 (1986): 470–491.

15. For an overview, see Sheldon Goldstein, "Bohmian Mechanics," *Stanford Encyclopedia of Philosophy* (Fall 2021 ed.), https://plato.stanford.edu/archives/fall2021/entries/qm-bohm/.

16. Carlo Rovelli, "Relational Quantum Mechanics," *International Journal of Theoretical Physics* 35 (1996): 1637–1678.

17. Rovelli, 1643.

18. See Don Howard, "Who Invented the 'Copenhagen Interpretation'? A Study in Mythology," *Philosophy of Science* 71 (2004): 669–682. See also Slobodan Perovic, *From Data to Quanta: Niels Bohr's Vision of Physics* (Chicago: University of Chicago Press, 2021).

19. Jan Faye, "Copenhagen Interpretation of Quantum Mechanics," *Stanford Encyclopedia of Philosophy* (Winter 2019 ed.), https://plato.stanford.edu/entries/qm-copenhagen/.

20. John von Neumann, *Mathematical Foundations of Quantum Mechanics*, new ed., ed. Nicholas A. Wheeler (Princeton, NJ: Princeton University Press, 2018). The original German text appeared in 1932.

21. Eugene Wigner, "Remarks on the Mind-Body Question," in *The Scientist Speculates: An Anthology of Partly-Baked Ideas*, ed. I. J. Good (New York: Basic Books, 1961), 171–184.

22. See C. A. Fuchs, "Notwithstanding Bohr, the Reasons for QBism," *Mind and Matter* 15 (2017): 245–300.

23. E. T. Jaynes, "Clearing Up Mysteries—the Original Goal," in *Maximum Entropy and Bayesian Methods*, ed. J. Skilling (Dordrecht: Kluwer, 1989), 7.

24. John B. DeBrota and Blake C. Stacey, "FAQBism," April 14, 2018, 1, https://arxiv.org/abs/1810.13401.

25. DeBrota and Stacey, 6.

26. N. David Mermin, "QBism Puts the Scientist Back into Science," *Nature* 507 (2014): 421–423.

27. N. David Mermin, "Why QBism Is Not the Copenhagen Interpretation and What John Bell Might Have Thought of It," in *Quantum [Un]Speakables II: Half a Century of Bell's Theorem*, ed. Reinhold Bertlmann and Anton Zeilinger (Cham: Switzerland: Springer, 2017), 88.

28. Thomas Nagel, *The View from Nowhere* (New York: Oxford University Press, 1986).

29. See John B. DeBrota, Christopher A. Fuchs, Jacques L. Pienaar, and Blake C. Stacey, "Born's Rule as a Quantum Extension of Bayesian Coherence," *Physical Review A* 104 (2021): 02207.

Chapter 5

1. See Jimena Canales, *The Physicist and the Philosopher: Einstein, Bergson, and the Debate That Changed Our Understanding of Time* (Princeton, NJ: Princeton University Press, 2015).

2. Canales.

3. Henri Bergson, *Duration and Simultaneity: Bergson and the Einsteinian Universe*, ed. Robin Durie, trans. Mark Lews and Robin Durie (Manchester: Clinamen Press, 1999).

4. For discussion of Bergson's errors, see Steven Savitt, "What Bergson Should Have Said to Einstein," *Bergsoniana* 1 (2021), https://doi.org/10.4000/bergsoniana.333. For discussion of Bergson's rigorous attention to the mathematical details of Einstein's theory, see Canales, *The Physicist and the Philosopher*, and C. S. Unnikrishnan, "The Theories of Relativity and Bergson's Philosophy of Duration and Simultaneity during and after Einstein's 1922 Visit to Paris" (2020), https://arxiv.org/abs/2001.10043.

5. C. S. Unnikrishnan, "Theories of Relativity," 4.

6. For a transcript of Bergson's remarks and Einstein's response, see Henri Bergson, "Remarks on the Theory of Relativity (1922)," *Journal of French and Francophone Philosophy—Revue de la philosophie française et de langue française* 28, no 1 (2020): 167–172. An earlier translation can be found in Bergson, *Duration and Simultaneity*, 154–159. For Bergson's celebrity status, see Emily Herring, "Henri Bergson, Celebrity," *Aeon*, May 6, 2019, https://aeon.co/essays/henri-bergson-the-philosopher -damned-for-his-female-fans.

7. Maurice Merleau-Ponty, "Einstein and the Crisis of Reason," in Maurice Merleau-Ponty, *Signs*, trans. Richard C. McCleary (Evanston, IL: Northwestern University Press, 1964), 195.

8. Bergson, *Duration and Simultaneity*, xxvii.

9. Bergson, "Remarks on the Theory of Relativity (1922)," 167.

10. Bergson, 170.

11. Bergson, 169.

12. Bergson, 171.

13. Bergson, 170.

14. Bertrand Russell, *A, B, C of Relativity* (New York: Routledge Press, 2009), 138.

15. Bergson, "Remarks on the Theory of Relativity (1922)," 170.

16. A. Einstein, "On the Electrodynamics of Moving Bodies," in H. A. Lorentz, A. Einstein, H. Minkowski, and H. Weyl, *The Principle of Relativity: A Collection of Original Memoirs on the Special and General Theory of Relativity*, trans. W. Perrett and G. B. Jeffrey (New York: Dover, 1952), 40.

17. Einstein, 40.

18. See William Lane Craig, "Bergson Was Right about Relativity (Well, Partly)!," in *Time and Tense: Unifying the Old and the New*, ed. Stamatios Gerogiorgakis (Munich: Philosophia Verlag, 2016), 317–352.

19. Einstein, "On the Electrodynamics of Moving Bodies," 39.

20. Bergson, "Remarks on the Theory of Relativity (1922)," 169, 170.

21. Durie, "Introduction," in Bergson, *Duration and Simultaneity*, xiv.

22. Durie, ix.

23. Henri Bergson, *Time and Free Will: An Essay on the Immediate Data of Consciousness*, trans. F. L. Pogson (Mineola, NY: Dover, 1913).

24. Henri Bergson, *Matter and Memory*, trans. Nancy Margaret Paul and W. Scott Palmer (New York: Zone Books, 1991).

25. Alfred North Whitehead, *The Concept of Nature* (Cambridge: Cambridge University Press, 1920; Ann Arbor: University of Michigan Press, 1957), 55. Citation refers to 1957 edition.

26. See Unnikrishnan, "Theories of Relativity."

27. For an accessible treatment of the twin paradox, see Paul Davies, *About Time: Einstein's Unfinished Revolution* (New York: Simon and Schuster, 1995), 59–67.

28. Bergson, *Duration and Simultaneity*, 145.

29. See Savitt, "What Bergson Should Have Said to Einstein," and Tim Maudlin, *Philosophy of Physics: Space and Time* (Princeton, NJ: Princeton University Press, 2021), 83.

30. J. C. Hafele and R. E. Keating, "Around-the-World Atomic Clocks: Predicted Relativistic Time Gains," *Science* 177 (1972): 166–168. For discussion, see Savitt, "What Bergson Should Have Said to Einstein," and Unnikrishnan, "Theories of Relativity."

31. See Jeremy Proulx, "Duration in Relativity: Some Thoughts on the Bergson-Einstein Encounter," *Southwest Philosophy Review* 32 (2017): 159.

32. See Craig, "Bergson Was Right."

33. Bergson, *Duration and Simultaneity*, 132.

34. Bergson, 123. See also Proulx, "Duration in Relativity," 160.

35. See Savitt, "What Bergson Should Have Said to Einstein"; Proulx, "Duration in Relativity"; and Peter Kügler, "What Bergson Should Have Said about Special Relativity," *Synthese* 198 (2021): 10273–10288.

36. N. David Mermin, "QBism as Cbism: Solving the Problem of 'the Now,'" December 30, 2013, https://arxiv.org/abs/1312.7825. See also N. David Mermin, "Making Better Sense of Quantum Mechanics," *Reports on Progress in Physics* 82 (2019): 012002, https://doi.org/10.1088/1361-6633/aae2c6.

37. Savitt, "What Bergson Should Have Said to Einstein," 95–96.

38. Proulx, "Duration in Relativity," 163.

39. Whitehead, *Concept of Nature*, 53.

40. For time to really behave like a spatial coordinate, the signs would have to be the same so that $ds^2 = + c^2(dt)^2 + (d\mathbf{X})^2$. This is a four-dimensional Euclidean space with coordinates ct and $\mathbf{X}$, which has no information about time flow, thus being very different from four-dimensional Minkowski space-time. In relativity theory, it's incorrect to say that time is just another spatial dimension, even though that's often repeated.

41. In a graphical representation of Minkowski space-time, the speed of light defines the light cone, a conic surface making a 45-degree angle with (one of) the space and the time coordinates. Physically allowed events, connected causally, live inside the light cone. Events outside the light cone, called "spacelike" events, are causally forbidden.

42. See Hans Reichenbach, *The Direction of Time* (New York: Dover, 1956), 11–12.

43. Hilary Putnam presents a classic argument for the block universe theory in "Time and Physical Geometry," *Journal of Philosophy* 64 (1967): 240–247. As Carlo Rovelli observes in *The Order of Time* (New York: Riverhead Books, 2018), 149–150, Putnam assumes that Einstein's definition of simultaneity has an ontological value, whereas it is mostly a definition of convenience for relating different times measured by different clocks. Our concern, however, is how we distinguish between mathematical abstraction and our perception of reality. By invoking the abstract mathematical construct of a four-dimensional space-time, the block universe theory does not solve the difficulties of understanding the obvious experiential fact that time flows.

44. Hermann Weyl, *Philosophy of Mathematics and Natural Science*, rev. and augmented English edition based on a translation by Olaf Helmer (Princeton, NJ: Princeton University Press, 1949), 116.

45. See Proulx, "Duration in Relativity," 153.

46. Bergson, *Duration and Simultaneity*, 107. Bergson's emphases.

47. See Proulx, "Duration in Relativity," 153.

48. *Albert Einstein—Michele Besso Correspondence, 1903–1955* (Paris: Hermann, 1972), 537–553.

49. See Rovelli, *The Order of Time*, 114–115.

50. For readers who are mathematically inclined, this example is worked out by Matthew Schwartz in his Statistical Mechanics notes available online at https://scholar.harvard.edu/files/schwartz/files/physics_181_lectures.pdf.

51. In quantum mechanics, position and momentum become Hermitian operators defined in Hilbert space. One can define expectation values of position and momentum, and those are the ones that matter for the uncertainty principle. The very concept of a precise location or momentum of a quantum object doesn't really make sense, adding to the idea of blurriness.

52. Lee Smolin argues this point effectively in his *Time Reborn: From the Crisis of Physics to the Future of the Universe* (Toronto: Vintage Canada, 2013).

53. For the curious, there are essentially three major contributions to the materials that fill the cosmos, each contributing differently to the expansion rate; whichever dominates determines the rate of expansion: radiation, or effectively massless relativistic particles (including photons and neutrinos); nonrelativistic massive particles (including weakly interacting dark matter); and an effective cosmological constant or vacuum energy, such as dark energy.

54. For the history of twentieth-century cosmology, see Marcelo Gleiser, *The Dancing Universe: From Creation Myths to the Big Bang* (Lebanon, NH: Dartmouth College Press, 2005).

55. For a critique of the idea of unification in physical theories, see Marcelo Gleiser, *A Tear at the Edge of Creation: A Radical New Vision for Life in an Imperfect Universe* (New York: Free Press, 2010); Peter Woit, *Not Even Wrong: The Failure of String Theory and the Continuing Challenge to Unify the Laws of Physics* (London: Jonathan Cape, 2006); Sabine Hossenfelder, *Lost in Math: How Beauty Leads Physics Astray* (New York: Basic Books, 2018); Lee Smolin, *Three Roads to Quantum Gravity: A New Understanding of Space, Time, and the Universe* (New York: Basic Books, 2001).

56. Isaiah Berlin, "Logical Translation," in Isaiah Berlin, *Concepts and Categories: Philosophical Essays*, ed. Henry Hardy (New York: Viking, 1979), 76.

57. See, for example, Erik Verlinde, "On the Origin of Gravity and the Laws of Newton," *Journal of High Energy Physics* 29 (2011), https://doi.org/10.1007/JHEP04 (2011)029.

58. We use "effectively massless" when the ratio of the particle's mass to the ambient radiation temperature is much smaller than one: $m/T \ll 1$.

59. Certain models of early-universe physics predict the existence of matter clumps of different kinds and masses, potentially formed after a period of very rapid expansion known as inflation. These include mini–black holes and oscillons, the latter codiscovered and named by one of us; see Marcelo Gleiser, "Pseudostable Bubbles," *Physical Review D* 49 (1994): 2978–2981. Although potentially important contributors to the total mass density of the universe, they would not dominate the expansion rate.

60. For the experts, this value assumes a flat universe with contributions of 4.8 percent normal matter, 26.2 percent dark matter, and 69 percent dark energy.

61. More precisely, photons also were affected by the gravitational pull from the dark matter wells and reacted by either falling in or escaping. These two opposite motions caused small changes in their frequencies called "acoustic oscillations," which are read today by detectors as small variations of 1 part in 100,000 in the smoothness of the cosmic microwave background temperature. These maps are absolutely crucial, as they allow us to reconstruct the conditions prevalent in the universe when it was only 400,000 years old.

62. David Albert, *Time and Chance* (Cambridge, MA: Harvard University Press, 2000).

63. Sean M. Carroll and Jennifer Chen, "Spontaneous Inflation and the Origin of the Arrow of Time" (2004), https://arxiv.org/abs/hep-th/0410270.

64. See Roberto M. Unger and Lee Smolin, *The Singular Universe and the Reality of Time* (Cambridge: Cambridge University Press, 2015), and Smolin, *Time Reborn*.

65. For a review, see, for example, A. Ijjas and Paul J. Steinhardt, "Bouncing Cosmology Made Simple," *Classical and Quantum Gravity* 35 (2018): 135004. For a different approach to bouncing cosmologies, see Stephon Alexander, Sam Cormack, and Marcelo Gleiser, "A Cyclic Approach to Fine Tuning," *Physics Letters B* 757 (2016): 247–250.

66. See Roger Jackson, "Dharmakīrti's Refutation of Theism," *Philosophy East and West* 36 (1986): 315–348, and Parimal G. Patil, *Against a Hindu God: Buddhist Philosophy of Religion in India* (New York: Columbia University Press, 2009).

Chapter 6

1. Georges Canguilhem, *Knowledge of Life*, trans. Stefanos Geroulanos and Daniela Ginsburg (New York: Fordham University Press, 2008), 90.

2. Hans Jonas, *The Phenomenon of Life: Toward a Philosophical Biology* (Evanston, IL: Northwestern University Press, 2001), 91.

3. Hans Jonas, "Is God a Mathematician? The Meaning of Metabolism," in Jonas, *The Phenomenon of Life*, 64–98.

4. Francisco J. Varela, *Principles of Biology Autonomy* (New York: Elsevier North-Holland, 1979); Alvaro Moreno and Matteo Mossio, *Biological Autonomy: A Philosophical and Theoretical Inquiry* (Dordrecht: Springer, 2015).

5. See Daniel J. Nicholson, "The Return of the Organism as a Fundamental Explanatory Concept in Biology," *Philosophy Compass* 9 (2014): 347–359; Tobias Cheung, "From the Organism of a Body to the Body of an Organism: Occurrence and Meaning of the Word 'Organism' from the Seventeenth to the Nineteenth Centuries," *British Journal for the History of Science* 39 (2006): 319–339.

6. For the Theoretical Biology Club, see Erik L. Peterson, *The Life Organic: The Theoretical Biology Club and the Roots of Epigenetics* (Pittsburgh, PA: University of Pittsburgh Press, 2017); Emily Herring and Gregory Radick, "Emergence in Biology: From Organicism to Systems Biology," in *The Routledge Handbook of Emergence*, ed. Sophie Gibb, Robin F. Hendry, and Tom Lancaster (London: Routledge Press, 2019), 352–362. For Barry Commoner, see his "Is DNA the 'Secret of Life'?," *Clinical Pharmacology and Therapeutics* 6 (1965): 273–278. For the return of the organism in late twentieth century and twenty-first century biology, see Nicholson, "The Return of the Organism." See also Evan Thompson, *Mind in Life: Biology, Phenomenology, and the Sciences of Mind* (Cambridge, MA: Harvard University Press, 2007), chaps. 5–7.

7. See Scott F. Gilbert and Sahotra Sarkar, "Embracing Complexity: Organicism for the 21st Century," *Developmental Dynamics* 219 (2000): 5.

8. Commoner, "Is DNA the 'Secret of Life'?," 276–277. See also Barry Commoner, "Unraveling the DNA Myth: The Spurious Foundation of Genetic Engineering," *Harper's Magazine* 304, no. 1821 (2002): 39–47.

9. Commoner, "Is DNA the 'Secret of Life'?," 273.

10. Nicholson, "The Return of the Organism," 353.

11. Erwin Schrödinger, *What Is Life? with Mind and Matter and Autobiographical Sketches* (Cambridge: Cambridge University Press, 1992).

12. Schrödinger, 125.

13. Eugene Wigner, "Physics and the Explanation of Life," *Foundations of Physics* 1, no. 1 (1970): 35–45.

14. Key reference points are Varela, *Principles of Biological Autonomy*; Robert Rosen, *Life Itself: A Comprehensive Inquiry into the Nature, Origin, and Fabrication of Life* (New York: Columbia University Press, 1991); Stuart A. Kauffman, *Investigations* (New York: Oxford University Press, 2000); Tibor Gánti, *The Principles of Life*, with commentary by James Gresemer and Eros Szathmary (Oxford: Oxford University Press, 2003); Moreno and Mossio, *Biological Autonomy*.

15. Immanuel Kant, *Critique of Judgment*, trans. W. S. Pluhar (Indianapolis, IN: Hackett, 1987).

16. Kant, 252–253. For discussion, see Evan Thompson, *Mind in Life: Biology, Phenomenology, and the Sciences of Mind* (Cambridge, MA: Harvard University Press, 2007), 129–140.

17. See Moreno and Mossio, *Biological Autonomy*; Matteo Mossio, Maël Montévil, and Giuseppe Longo, "Theoretical Principles for Biology: Organization," *Progress in Biophysics and Molecular Biology* 122 (2016): 24–35.

18. See Juan Carlos Letelier, Maria Luz Cárdenas, and Athel Cornish-Bowden, "From *L'Homme Machine* to Metabolic Closure: Steps towards Understanding Life," *Journal of Theoretical Biology* 286 (2011): 100–113.

19. Jean Piaget, *Biology and Knowledge: An Essay on the Relations between Organic Regulations and Cognitive Processes* (Chicago: University of Chicago Press, 1971).

20. Piaget, 155–156.

21. Varela, *Principles of Biological Autonomy*, 55–58.

22. Francisco J. Varela, "The Creative Cycle: Sketches on the Natural History of Circularity," in *The Invented Reality*, ed. Paul Watzlavick (New York: Norton, 1984), 311–312.

23. Humberto R. Maturana and Francisco J. Varela, *Autopoiesis and Cognition: The Realization of the Living* (Dordrecht: D. Reidel, 1980).

24. Rosen, *Life Itself*.

25. Stuart A. Kauffman, "Cellular Homeostasis, Epigenesis, and Replication in Randomly Aggregated Macromolecular Systems," *Journal of Cybernetics* 1 (1971): 71–96; Stuart A. Kauffman, "Autocatalytic Sets of Proteins," *Journal of Theoretical Biology* 119 (1986): 1–24.

26. Kauffman, *Investigations*.

27. Moreno and Mossio, *Biological Autonomy*; Mossio et al., "Theoretical Principles for Biology: Organization."

28. Mossio et al., "Theoretical Principles for Biology: Organization," 7.

29. Mossio et al., 7.

30. See Moreno and Mossio, *Biological Autonomy*, chap. 3.

31. See Fermin Fulda, "Natural Agency: The Case of Bacterial Cognition," *Journal of the American Philosophical Association* 3 (2017): 69–90.

32. Moreno and Mossio, *Biological Autonomy*, 98.

33. See Ezequiel Di Paolo, "Autopoiesis, Adaptivity, Teleology, Agency," *Phenomenology and the Cognitive Sciences* 4 (2005): 429–452.

34. Francisco J. Varela, "Living Ways of Sense-Making: A Middle Path for Neuroscience," in *Disorder and Order: Proceedings of the Stanford International Symposium (September 14–16, 1981)*, ed. Paisley Livingston (Saratoga, CA: Anma Libri, 1984), 208–224; Francisco J. Varela, "Patterns of Life: Intertwining Identity and Cognition," *Brain and Cognition* 34 (1997): 72–87.

35. Artemy Kolchinsky and David H. Wolpert, "Semantic Information, Autonomous Agency and Non-Equilibrium Statistical Physics," *Interface Focus* 8 (2018): 20180041, http://dx.doi.org/10.1098/rsfs.2018.0041.

36. Evan Thompson, "Living Ways of Sense-Making," *Philosophy Today*, suppl. (2011): 114–123.

37. See Di Paolo, "Autopoiesis, Adaptivity, Teleology, Agency," and Ezequiel Di Paolo and Evan Thompson, "The Enactive Approach," in *The Routledge Handbook of Embodied Cognition*, ed. Lawrence Shapiro (London: Routledge Press, 2014), 68–78.

38. Ezequiel Di Paolo, "Extended Life," *Topoi* 28 (2009): 16n5. See also Ezequiel Di Paolo, "The Enactive Conception of Life," in *The Oxford Handbook of 4E Cognition*, ed. Albert Newen, Leon De Bruin, and Shaun Gallagher (Oxford: Oxford University Press, 2018), 71–94.

39. Theodosius Dobzhansky, "Nothing in Biology Makes Sense Except in the Light of Evolution," *American Biology Teacher* 35 (1973): 125–129.

40. Moreno and Mossio, *Biological Autonomy*, 113.

41. Moreno and Mossio, 116. See also Thompson, *Mind in Life*, 167–170.

42. Marcelo Gleiser and Sara Imari Walker, "Toward Homochiral Protocells in Non-catalytic Peptide Systems," *Origins of Life and Evolution of the Biosphere* 39 (2009): 479–493; see also Pier Luigi Luisi, *The Emergence of Life: From Chemical Origins to Synthetic Biology*, 2nd ed. (Cambridge: Cambridge University Press, 2019).

43. Moreno and Mossio, *Biological Autonomy*, 135–137. See also Kepa Ruiz-Mirazo, Jon Umerez, and Alvaro Moreno, "Enabling Conditions for 'Open-Ended Evolution,'" *Biology and Philosophy* 23 (2008): 67–85.

44. Kepa Ruiz-Mirazo, Juli Peretó, and Alvaro Moreno, "A Universal Definition of Life: Autonomy and Open-Ended Evolution," *Origins of Life and Evolution of the Biosphere* 34 (2004): 323–346.

45. See Daniel J. Nicholson, "Organisms ≠ Machines," *Studies in History and Philosophy of Biological and Biomedical Sciences* 44 (2013): 669–678, and Daniel J. Nicholson, "Is the Cell *Really* a Machine?," *Journal of Theoretical Biology* 477 (2019): 108–126.

46. The concepts of decomposable, nearly decomposable, and nondecomposable come from Herbert Simon, *The Sciences of the Artificial* (Cambridge, MA: MIT Press, 1969).

47. Nicholson, "Is the Cell *Really* a Machine?"

48. Nicholson, 110.

49. Nicholson, 123.

50. Nicholson, 122.

51. Stuart Kauffman, *A World beyond Physics: The Emergence and Evolution of Life* (New York: Oxford University Press, 2019).

52. Giuseppe Longo, Maël Montévil, and Stuart Kauffman, "No Entailing Laws, But Enablement in the Evolution of the Biosphere," *Proceedings of the 14th Annual Conference Companion on Genetic and Evolutionary Computation* (New York: ACM, 2012), 1379–1392, https://dl.acm.org/doi/10.1145/2330784.2330946; Giuseppe Longo and Maël Montévil, "Extended Criticality, Phase Spaces, and Enablement in Biology," *Chaos, Solitons and Fractals* 55 (2013): 64–79. See also Giuseppe Longo and Maël Montévil, *Perspectives on Organisms: Biological Time, Symmetries and Singularities* (Berlin: Springer-Verlag, 2014); Kauffman, *A World beyond Physics*.

53. Longo et al., "No Entailing Laws," 1390.

54. See David D. Nolte, "The Tangled Tale of Phase Space," *Physics Today* 63 (2010): 33–38.

55. Longo et al., "No Entailing Laws," 1384.

56. Stephen Jay Gould and Elisabeth S. Vrba, "Exaptation—a Missing Term in the Science of Form," *Paleobiology* 8 (1982): 4–15.

57. Longo et al., "No Entailing Laws."

58. Stuart Kauffman, "Answering Schrödinger's Question, 'What Is Life?,'" *Entropy* 22 (2020): 8, https://doi.org/10.3390/e22080815.

59. This point raises the philosophical issue of emergence. Kauffman's idea that life has no prestateable phase space is an example of what philosopher Paul Humphreys calls "diachronic emergence" (emergence over time). There is now a large technical literature on emergence in the philosophy of science and metaphysics. We have chosen not to engage this literature here because it goes beyond our critique of the Blind Spot. See Paul Humphreys, *Emergence: A Philosophical Account* (New York: Oxford University Press, 2016); Jessica M. Wilson, *Metaphysical Emergence* (New York: Oxford University Press, 2021); Robert C. Bishop, Michael Silberstein, and Mark Pexton, *Emergence in Context: A Treatise in Twenty-First Century Natural Philosophy* (New York: Oxford University Press, 2022).

60. Stuart Kauffman, "Is There a Fourth Law for Non-Ergodic Systems That Do Work to Construct Their Expanding Phase Space?" (2022), https://doi.org/10.48550/arXiv.2205.09762.

61. Francisco J. Varela, "Laying Down a Path in Walking," in *Gaia: A Way of Knowing: Political Implications of the New Biology*, ed. William Irwin Thompson (Hudson, NY: Lindisfarne Press, 1987), 48–64.

62. See Longo and Montévil, *Perspectives on Organisms*; Ana M. Soto, Giuseppe Longo, Paul-Antoine Miquel, Maël Montévil, Matteo Mossio, Nicole Perret, Arnaud Pocheville, and Carlos Sonnenschein, "Toward a Theory of Organisms: Three Founding Principles in Search of a Useful Integration," *Progress in Biophysics and Molecular Biology* 122 (2016): 77–82; D. M. Walsh, *Organisms, Agency, and Evolution* (Cambridge: Cambridge University Press, 2015).

63. Kauffman, *A World Beyond Physics*, 128.

Chapter 7

1. See Francisco Varela, Evan Thompson, and Eleanor Rosch, *The Embodied Mind: Cognitive Science and Human Experience*, rev. ed. (2017; repr., Cambridge, MA: MIT Press, 1991).

2. Varela et al., *The Embodied Mind*. See also Evan Thompson, *Mind in Life: Biology, Phenomenology, and the Sciences of Mind* (Cambridge, MA: Harvard University Press, 2007); Ezequiel Di Paolo, Elena Clare Cuffari, and Hanne De Jaegher, *Linguistic Bodies: The Continuity between Life and Language* (Cambridge, MA: MIT Press, 2018).

3. See John Vervaeke, Timothy Lillicrap, and Blake A. Richards, "Relevance Realization and the Emerging Framework in Cognitive Science," *Journal of Logic and Computation* 22, no. 1 (2012): 79–99.

4. We take the term *relevance realization* from Vervaeke et al., "Relevance Realization."

5. See Brian Cantwell Smith, *The Promise of Artificial Intelligence: Reckoning and Judgment* (Cambridge, MA: MIT Press, 2019). Melanie Mitchell, *Artificial Intelligence: A Guide for Thinking Humans* (New York: Farrar, Straus and Giroux, 2019). Erik J. Larson, *The Myth of Artificial Intelligence: Why Computers Can't Think the Way We Do* (Cambridge, MA: Harvard University Press, 2021). For the classic yet still highly relevant critique of AI, see Hubert Dreyfus, *What Computers Can't Do: A Critique of Artificial Reason* (New York: Harper & Row, 1972), and *What Computers Still Can't Do: A Critique of Artificial Reason* (Cambridge, MA: MIT Press, 1992).

6. John Searle, *Minds, Brains and Science* (Cambridge, MA: Harvard University Press, 1986).

7. See Mitchell, *Artificial Intelligence*, and Larson, *The Myth of Artificial Intelligence*.

8. See Mitchell, *Artificial Intelligence*, 110–116.

9. Smith, *The Promise of Artificial Intelligence*.

10. For an overview, see Murray Shanahan, "The Frame Problem," *Stanford Encyclopedia of Philosophy*, last revised February 8, 2016, https://plato.stanford.edu/archives/spr2016/entries/frame-problem/.

11. Daniel C. Dennett, "Cognitive Wheels: The Frame Problem of AI," in *Minds, Machines and Evolution: Philosophical Studies*, ed. Christopher Hookway (Cambridge: Cambridge University Press, 1984), 129–151.

12. This is true for both the narrowly defined, technical version of the frame problem and the wider version known as the generalized frame problem or the epistemological frame problem. See Shanahan, "The Frame Problem." The frame problem we have described is the generalized version.

13. See Mitchell, *Artificial Intelligence*, chap. 9.

14. David Silver, Julian Schrittwieser, Karen Simonyan, Ioannis Antonoglou, Aja Huang, Arthur Guez, Thomas Hubert, et al., "Mastering the Game of Go without Human Knowledge," *Nature* 550 (2017): 354–359; Michael Nielsen, "Is AlphaGo Really Such a Big Deal?," *Quanta Magazine*, March 29, 2016.

15. Silver et al., "Mastering the Game of Go."

16. Gary Marcus and Ernest Davis, *Rebooting AI: Building Artificial Intelligence We Can Trust* (New York: Pantheon Books, 2019), 145–146. See also Larson, *The Myth of Artificial Intelligence*, 161–162.

17. Smith, *The Promise of Artificial Intelligence*, 77–78.

18. Contrary to Nielsen, "Is AlphaGo Really Such a Big Deal?"

19. Mitchell, *Artificial Intelligence*, 166.

20. John Pavlus, "The Computer Scientist Training AI to Think with Analogies" (Interview with Melanie Mitchell), *Quanta Magazine*, July 14, 2021, https://www.quantamagazine.org/melanie-mitchell-trains-ai-to-think-with-analogies-20210714/#.

21. See Vervaeke et al., "Relevance Realization."

22. Mitchell, *Artificial Intelligence*, 166–167, 256. For an example of recent advances on open-ended learning, see DeepMind's "Open-Ended Learning Leads to Generally Capable Agents," July 27, 2021, https://www.deepmind.com/publications/open-ended-learning-leads-to-generally-capable-agents.

23. Mitchell, *Artificial Intelligence*, 168–169. See also Gary Marcus, "Innateness, AlphaZero, and Artificial Intelligence" (2018), arXiv:1801.05667.

24. Mitchell, *Artificial Intelligence*, 172.

25. Mitchell, 268.

26. See Mitchell, 270.

27. Smith, *The Promise of Artificial Intelligence*, 7–8.

28. Hubert Dreyfus, "Intelligence without Representation: Merleau-Ponty's Critique of Mental Representation," *Phenomenology and the Cognitive Sciences* 1, no. 4 (2002): 413–425.

29. Smith, *The Promise of Artificial Intelligence*, 24, 28, 38, 67.

30. Stephen Grossberg, "Adaptive Resonance Theory: How a Brain Learns to Consciously Attend, Learn, and Recognize a Changing World," *Neural Networks* 37 (2013): 1–47. See also Stephen Grossberg, *Conscious Mind, Resonant Brain: How Each Brain Makes a Mind* (New York: Oxford University Press, 2021).

31. Smith, *The Promise of Artificial Intelligence*, xiv.

32. Smith, 35.

33. See Mitchell, *Artificial Intelligence*, 106–108, 195–196; Kate Crawford, *Atlas of AI: Power, Politics, and the Planetary Crisis of AI* (New Haven: Yale University Press, 2021), chap. 4; Abeba Birhane, "The Impossibility of Automating Ambiguity," *Artificial Life* 27 (2021): 44–61.

34. Birhane, "The Impossibility of Automating Ambiguity," 44.

35. Mitchell, *Artificial Intelligence*, 106–107.

36. Smith, *The Promise of Artificial Intelligence*, 50.

37. See Melanie Mitchell and David C. Krakauer, "The Debate Over Understanding in AI's Large Language Models," *Proceedings of the National Academy of Sciences* 120, no. 13 (2023): e2215907120.

38. Mitchell, *Artificial Intelligence*, 104–105.

39. Abeba Birhane, Pratyusha Kalluri, Dallas Card, William Agnew, Ravit Dotan, and Michelle Bao, "The Values Encoded in Machine Learning Research," June 21, 2022, arXiv:2016.15590 [cs.LG].

40. Crawford, *Atlas of AI*, 135–136.

41. Crawford, 8.

42. Crawford, 15.

43. Crawford, 8.

44. See Varela et al., *The Embodied Mind*; Thompson, *Mind in Life*.

45. Di Paolo et al., *Linguistic Bodies*, 21.

46. Stephen Grossberg, "A Path toward Explainable AI and Autonomous Adaptive Intelligence: Deep Learning, Adaptive Resonance, and Models of Perception, Emotion, and Action," *Frontiers in Neurorobotics*, June 25, 2020, doi.org/10.3389/fnbot .2020.00036.

47. Crawford, *Atlas of AI*, 8.

48. Smith, *The Promise of Artificial Intelligence*, xix–xx.

49. Smith, xx.

Chapter 8

1. See D. E. Harding, *On Having No Head: Zen and the Rediscovery of the Obvious* (London: Buddhist Society, 1961; London: Sholland Trust, 2014). See also Brentyn J. Ramm, "Pure Awareness Experience," *Inquiry* (2019), doi:10.1080/0020174X.2019.1592.

2. Ludwig Wittgenstein, *Tractatus Logico-Philosophicus*, trans. D. F. Pears and B. F. McGuinness (London: Routledge Classics, 2001), 5.633, 69.

3. The term *aware-space* comes from Harding, *On Having No Head*. See also Ramm, "Pure Awareness Experience."

4. See Maurice Merleau-Ponty, *Signs*, trans. Richard C. McCleary (Evanston, IL: Northwestern University Press, 1964), 166.

5. Harding, *On Having No Head*.

6. See Gilbert Harman, "The Intrinsic Quality of Experience," in *The Nature of Consciousness: Philosophical Debates*, ed. Ned Block, Owen Flanagan, and Guven Güzeldere (Cambridge, MA: MIT Press, 1997), 663–676.

7. See Thomas Metzinger, *Being No One: The Self-Model Theory of Subjectivity* (Cambridge, MA: MIT Press, 2003), 163–179.

8. See Evan Thompson, *Mind in Life: Biology, Phenomenology, and the Sciences of Mind* (Cambridge, MA: Harvard University Press, 2007), 282–287.

9. See Evan Thompson, *Waking, Dreaming, Being: Self and Consciousness in Neuroscience, Meditation, and Philosophy* (New York: Columbia University Press, 2015), chap. 5.

10. The concept of "witness consciousness" comes from Indian philosophy. We are using it here strictly in a phenomenological sense, without commitment to the larger background metaphysics in which it has traditionally figured, according to which the only ultimate reality is consciousness. See Bina Gupta, *The Disinterested Witness: A Fragment of Advaita Vedānta Phenomenology* (Evanston, IL: Northwestern

University Press, 1998). See also Miri Albahari, "Witness-Consciousness: Its Definition, Appearance, and Reality," *Journal of Consciousness Studies* 16, no. 1 (2009): 62–84.

11. See John D. Dunne, Evan Thompson, and Jonathan Schooler, "Mindful Meta-Awareness: Sustained and Non-Propositional," *Current Opinion in Psychology* 28 (2019): 307–311. See also John D. Dunne, "Buddhist Styles of Mindfulness: A Heuristic Approach," in *Handbook of Mindfulness and Self-Regulation*, ed. Brian D. Ostafin, Michael D. Robinson, and Brian Meier (New York: Springer, 2015), 251–270.

12. The subtitle of Harding's *On Having No Head* is *Zen and the Rediscovery of the Obvious*. See also Brentyn Ramm, "The Technology of Awakening: Experiments in Zen Phenomenology," *Religions* 12, no. 3 (2021): 192, https://doi.org/10.3390/rel12030192.

13. For an introduction to Dōgen's life and thought, see Hee-Jin Kim, *Eihei Dōgen: Mystical Realist* (Boston: Wisdom Publications, 2004).

14. Robert H. Sharf, "Epilogue: Mind in World, World in Mind," in Sharf, *What Can't Be Said: Paradox and Contradiction in East Asian Thought*, ed. Yasuo Deguchi, Jay L. Garfield, Graham Priest, and Robert H. Sharf (New York: Oxford University Press, 2021), 192.

15. We take the terms *closure* and *transcendence* from Graham Priest, *Beyond the Limits of Thought* (Oxford: Oxford University Press, 2002).

16. See Edmund Husserl, *The Idea of Phenomenology*, trans. Lee Hardy (Dordrecht: Kluwer Academic, 1999), 23, 33, 37; Edmund Husserl, *Ideas: General Introduction to Pure Phenomenology*, trans. W. R. Boyce Gibson (New York: Macmillan, 1931; London: Routledge, 2002), sec. 32. Citations refer to the 2002 edition.

17. The term *spiritual exercise* comes from Pierre Hadot, *Philosophy as a Way of Life*, trans. Michael Chase (Oxford: Blackwell, 1995).

18. See Michel Bitbol, "Is Consciousness Primary?," *NeuroQuantology* 6 (2008): 53–72.

19. For more on the horizonal conception of consciousness, see J. J. Valberg, *Dream, Death, and the Self* (Princeton, NJ: Princeton University Press, 2007). The term *horizon* in this context comes from Husserl. See Salius Geniusas, *The Origins of the Horizon in Husserl's Phenomenology* (New York: Springer, 2012).

20. See Bitbol, "Is Consciousness Primary?"

21. Maurice Merleau-Ponty, *Phenomenology of Perception*, trans. D. Landes (London: Routledge Press, 2013), 84.

22. We take the term *strange loop* from Douglas Hofstadter. See Douglas Hofstadter, *Gödel, Escher, Bach: An Eternal Golden Braid* (New York: Basic Books, 1979), and

Douglas Hofstadter, *I Am a Strange Loop* (New York: Basic Books, 2007). See also Francisco J. Varela, "The Creative Circle: Sketches on the Natural History of Circularity," in *The Invented Reality*, ed. Paul Watzlavick (New York: Norton, 1984), 309–323; and Sharf, "Epilogue: Mind in World, World in Mind."

23. Merleau-Ponty, *Phenomenology of Perception*, 454.

24. See Ezequiel Di Paolo, "The Enactive Conception of Life," in *The Oxford Handbook of 4E Cognition*, ed. Alfred Newen, Leon De Bruin, and Shaun Gallagher (Oxford: Oxford University Press), 71–94.

25. Merleau-Ponty, *Phenomenology of Perception*, 456.

26. See David Suarez, "Nature at the Limits of Science and Phenomenology," *Journal of Transcendental Philosophy* 1, no. 1 (2020): 109–133.

27. Philosopher David Chalmers coined the term "the hard problem of consciousness" in the 1990s. See David J. Chalmers, *The Conscious Mind: In Search of a Fundamental Theory* (New York: Oxford University Press, 1996).

28. Thomas H. Huxley, *Lessons in Elementary Physiology* (London: Macmillan, 1866), 193.

29. John Tyndall, as quoted by William James, *The Principles of Psychology* (Cambridge, MA: Harvard University Press, 1981), 150. See also Marcelo Gleiser, *The Island of Knowledge: The Limits of Science and the Search for Meaning* (New York: Basic Books, 2014), 266–267.

30. The term *explanatory gap* comes from philosopher Joseph Levine. See Joseph Levine, "Materialism and Qualia: The Explanatory Gap," *Pacific Philosophical Quarterly* 64, no. 4 (1983): 354–361.

31. John Locke, *An Essay Concerning Human Understanding*, ed. Roger Woolhouse (London: Penguin, 1997), bk. 4, chap. 3, sec. 13, 484.

32. G. W. Leibniz, *Philosophical Essays*, ed. and trans. Roger Ariew and Daniel Garber (Indianapolis: Hackett, 1989), 215.

33. See Thomas Nagel, "What Is It Like to Be a Bat?," *Philosophical Review* 83 (1974): 435–450, and Chalmers, *Conscious Mind*.

34. See Bitbol, "Is Consciousness Primary?"

35. For naturalistic dualism, see Chalmers, *Conscious Mind*, and David Chalmers, "Facing Up to the Hard Problem of Consciousness," *Journal of Consciousness Studies* 2, no. 2 (1995): 200–219. For panpsychism and its relationship to naturalistic dualism, see William Seager, "Consciousness, Information, and Panpsychism," *Journal of Consciousness Studies* 2, no. 3 (1995): 272–288.

36. The term "extra ingredient" comes from Chalmers. See Chalmers, "Facing Up to the Hard Problem of Consciousness."

37. For a recent overview, see Anil K. Seth and Tim Bayne, "Theories of Consciousness," *Nature Reviews Neuroscience* (2022), https://doi.org/10.1038/s41583-022-00587-4.

38. See Christof Koch, Marcello Massimini, Melanie Boly, and Giulio Tononi, "Neural Correlates of Consciousness: Progress and Problems," *Nature Reviews Neuroscience* 17 (2016): 307–321.

39. Koch et al., 307–321.

40. See Francesca Siclari, Benjamin Baird, Lampros Perogamvros, Giulio Bernardi, Joshua J. LaRocque, Brady Riedner, Melanie Boly, Bradley R. Postle, and Giulio Tononi, "The Neural Correlates of Dreaming," *Nature Neuroscience* 20 (2017): 872–878. See also Jennifer M. Windt, Tore Nielsen, and Evan Thompson, "Does Consciousness Disappear in Dreamless Sleep?," *Trends in Cognitive Sciences* 20 (2016): 871–882.

41. See Melanie Boly, Marcello Massimini, Naotsugu Tsuchiya, Bradley R. Postle, Christof Koch, and Giulio Tononi, "Are the Neural Correlates of Consciousness in the Front or in the Back of the Cerebral Cortex? Clinical and Neuroimaging Evidence," *Journal of Neuroscience* 37 (2017): 9603–9613, and Brian Odegard, Robert T. Knight, and Hakwan Lau, "Should a Few Null Findings Falsify Prefrontal Theories of Conscious Perception?," *Journal of Neuroscience* 37 (2017): 9593–9602.

42. See Ned Block, "What Is Wrong with the No-Report Paradigm and How to Fix It," *Trends in Cognitive Sciences* 23 (2019): 1003–1013.

43. See Evan Thompson, *Waking, Dreaming, Being: Self and Consciousness in Neuroscience, Meditation, and Philosophy* (New York: Columbia University Press, 2015). See also Thomas Metzinger, *The Ego Tunnel: The Science of the Mind and the Myth of the Self* (New York: Basic Books, 2010), and Anil Seth, *Being You: A New Science of Consciousness* (New York: Dutton, 2021).

44. Seth, for example, makes this claim in *Being You*, 30–31.

45. Bitbol, "Is Consciousness Primary?"

46. See Evan Thompson, "Could All Life Be Sentient?," *Journal of Consciousness Studies* 29 (2022): 229–265.

47. For restricting consciousness to animals with brains, see Simona Ginsburg and Eva Jablonka, *The Evolution of the Sensitive Soul: Learning and the Origins of Consciousness* (Cambridge, MA: MIT Press, 2019); Peter Godfrey-Smith, *Metazoa: Animal Life and the Birth of the Mind* (New York: Farrar, Straus, and Giroux, 2020). For a case

that all life is sentient, see Arthur Reber, *The First Minds: Caterpillars, 'Karyotes, and Consciousness* (New York: Oxford University Press, 2019). For critical discussion of these options, see Thompson, "Could All Life Be Sentient?"

48. See Eric Schwitzgebel, "Is There Something It's Like to Be a Garden Snail?" (2020), http://www.faculty.ucr.edu/~eschwitz/SchwitzPapers/Snails-201223.pdf.

49. See, for example, Seth, *Being You*, 31–33.

50. See Thompson, *Mind in Life*.

51. See Seth, *Being You*.

52. See Jakob Howy, *The Predictive Mind* (New York: Oxford University Press, 2013).

53. Seth, *Being You*, 87–89, 111–112. Seth credits psychologist Chris Frith for the phrase "controlled hallucination." Neuroscientist Rodolfo Llínas stated a similar idea when he said that perception is a dreamlike state constrained by the senses. See Rodolfo Llínas, "Perception as an Oneiric State Modulated by the Senses," in *Large-Scale Neuronal Theories of the Brain*, ed. Christof Koch and Joel L. Davis (Cambridge, MA: MIT Press, 1994), 111–124. See also Aaron Sloman, "Experiencing Computation: A Tribute to Max Clowes," April 2014, https://www.cs.bham.ac.uk/research/projects/cogaff/sloman-clowestribute.html; and Aaron Sloman, "What the Brain's Mind Tells the Mind's Eye," November 29, 2005, https://www.cs.bham.ac.uk/research/projects/cogaff/sloman-vis-affordances.pdf.

54. Llínas, "Perception as an Oneiric State Modulated by the Senses."

55. For problems with the predictive processing theory, see Kevin S. Walsh, David McGovern, Andy Clark, and Redmond G. O'Connell, "Evaluating the Neurophysiological Evidence for Predictive Processing as a Model of Perception," *Annals of the New York Academy of Sciences* 1464, no. 1 (2020): 242–268; Johan Kwisthout and Iris van Rooij, "Computational Resource Demands of a Predictive Bayesian Brain," *Computational Brain and Behavior* 3 (2020): 174–188; Madeleine Ransom, Sina Fazelpour, Jelena Markovic, James Kryklywy, Evan T. Thompson, and Rebecca M. Todd, "Affect-Biased Attention and Predictive Processing," *Cognition* 203 (2020), https://doi.org/10.1016/j.cognition.2020.104370.

56. Luiz Pessoa (@PessoaBrain), "No one calls the brain the **Newtonian Brain**," Twitter, March 26, 2022, 1:04 p.m., https://twitter.com/PessoaBrain/status/1507765399391195138.

57. The phrase "the math is not the territory" comes from Mel Andrews, "The Math Is Not the Territory: Navigating the Free Energy Principle," *Biology and Philosophy* 30 (2021), https://doi.org/10.1007/s10539-021-09807-0.

58. See Jakob Hohwy and Anil Seth, "Predictive Processing as a Systematic Basis for Identifying the Neural Correlates of Consciousness," *Philosophy and the Mind Sciences* (2020), https://doi.org/10.33735/phimisci.2020.II.64.

59. The term "generation problem" for consciousness comes from William Seager, *Metaphysics of Consciousness* (London: Routledge Press, 1991).

60. Seth, *Being You*, 120; see also 80; Anil K. Seth, "The Real Problem," *Aeon*, November 2, 2016, https://aeon.co/essays/the-hard-problem-of-consciousness-is-a-distraction-from-the-real-one.

61. J. J. Gibson, *The Ecological Approach to Visual Perception* (Hillsdale, NJ: Erlbaum, 1979). See also Anthony P. Chemero, *Radical Embodied Cognitive Science* (Cambridge, MA: MIT Press, 2009).

62. See Alva Noë, *Action in Perception* (Cambridge, MA: MIT Press, 2004); Ezequiel Di Paolo, Thomas Buhrmann, and Xabier Barandiaran, *Sensorimotor Life: An Enactive Proposal* (Oxford: Oxford University Press, 2017).

63. See Alva Noë, "Review of Andy Clark, *Surfing Uncertainty: Prediction, Action, and the Embodied Mind*," *Mind* 127 (2018): 611–618.

64. See Jan Westerhoff, *The Non-Existence of the Real World* (Oxford: Oxford University Press, 2020), 66–68.

65. We take this analogy from Evan Thompson, *Why I Am Not a Buddhist* (New Haven: Yale University Press, 2020), 123.

66. This image comes from Jakob Hohwy, *The Predictive Mind* (Oxford: Oxford University Press, 2013), 15–16.

67. See Alva Noë, *Out of Our Heads: Why You Are Not Your Brain and Other Lessons from the Biology of Consciousness* (New York: Hill and Wang, 2009); Evan Thompson and Diego Cosmelli, "Brain in a Vat or Body in a World? Brainbound versus Enactive Views of Experience," *Philosophical Topics* 39 (2011): 163–180; and Thomas Fuchs, *Ecology of the Brain: The Phenomenology and Biology of the Embodied Mind* (Oxford: Oxford University Press, 2018).

68. See Giulio Tononi, "Integrated Information Theory," *Scholarpedia* 10, no. 1 (2015): 4164, doi:10.4249/scholarpedia.4164.

69. Giulio Tononi, "Consciousness as Integrated Information: A Provisional Manifesto," *Biological Bulletin* 215 (2008): 216–242.

70. Tononi, 216–242.

71. Giulio Tononi, Melanie Boly, Marcello Massimini, and Christof Koch, "Integrated Information Theory: From Consciousness to Its Physical Substrate," *Nature Reviews Neuroscience* 17 (2016): 450–461.

72. Erik P. Hoel, Larissa Albantakis, and Giulio Tononi, "Quantifying Causal Emergence Shows that Macro Can Beat Micro," *Proceedings of the National Academy of Sciences* 110 (2013): 19790–19795; Erik P. Hoel, Larissa Albantakis, William Marshall, and Giulio Tononi, "Can the Macro Beat the Micro? Integrated Information Across Spatiotemporal Scales," *Neuroscience of Consciousness* 1 (2016): 1–13.

73. Miguel Aguilera and Ezequiel A. Di Paolo, "Integrated Information in the Thermodynamic Limit," *Neural Networks* 114 (2019): 136–146.

74. See Husserl, *Ideas*, sec. 73–74.

75. See Tim Bayne, "On the Axiomatic Foundations of the Integrated Information Theory of Consciousness," *Neuroscience of Consciousness* 4 (2018): 1–8.

76. See Jay Garfield, *The Fundamental Wisdom of the Middle Way: Nāgārjuna's Mūlamadhyamakakārikā* (New York: Oxford University Press, 1995), 89–90, 220–224; Jan Westerhoff, *Nāgārjuna's Madhyamaka: A Philosophical Investigation* (Oxford: Oxford University Press, 2009), chap. 2.

77. See Thompson, *Waking, Dreaming, Being*, chaps. 3 and 8. See also Thomas Metzinger, "Minimal Phenomenal Experience: Meditation, Tonic Alertness, and the Phenomenology of 'Pure' Consciousness," *Philosophy and the Mind Sciences* 1 (2020), https://doi.org/10.33735/phimisci.2020.I.46.

78. See Tim Bayne, "Unity of Consciousness," *Scholarpedia* 4, no. 2 (2009): 7414, doi:10.4249/scholarpedia.7414.

79. See Bayne, "On the Axiomatic Foundations." See also Bjorn Merker, Kenneth Williford, and David Rudrauf, "The Integrated Information Theory of Consciousness: A Case of Mistaken Identity," *Behavioral and Brain Sciences* 45 (2022): E41, doi:10.1017/S0140525X21000881.

80. For further criticisms along these lines, see Merker et al., "The Integrated Information Theory of Consciousness."

81. For discussion of this point in relation to IIT, see Mike Beaton and Igor Aleksander, "World-Related Integrated Information: Enactivist and Phenomenal Perspectives," *International Journal of Machine Consciousness* 4 (2012): 439–455.

82. See Giulio Tononi and Christof Koch, "Consciousness: Here, There, and Everywhere?," *Philosophical Transactions of the Royal Society B* 370 (2015): 201460167, https://doi.org/10.1098/rstb.2014.0167; Hedda Hassel Mørch, "Is the Integrated Information Theory of Consciousness Compatible with Russellian Panpsychism?," *Erkenntnis* 84 (2018): 1065–1085. For criticism of IIT on the grounds that it leads to panpsychism, see Merker et al., "The Integrated Information Theory of Consciousness."

83. See Marcelo Gleiser, *The Island of Knowledge: The Limits of Science and the Search for Meaning* (New York: Basic Books, 2014).

84. Carl G. Hempel, "Comments on Goodman's Ways of Worldmaking," *Synthese* 45 (1980): 193–199. See also Tim Crane and D. H. Mellor, "There Is No Question of Physicalism," *Mind* 99 (1990): 185–206, and Noam Chomsky, *New Horizons in the Study of Language and Mind* (Cambridge: Cambridge University Press, 2000).

85. Henry Stapp, "Quantum Approaches to Consciousness," in *The Cambridge Handbook of Consciousness*, ed. Philip David Zelazo, Morris Moscovitch, and Evan Thompson (New York: Cambridge University Press, 2007), 881–907.

86. Galen Strawson, "Realistic Monism—Why Physicalism Entails Panpsychism," *Journal of Consciousness Studies* 13, nos. 10–11 (2006): 3–31.

87. Galen Strawson, "The Consciousness Deniers," *New York Review of Books*, March 13, 2018.

88. Versions of this argument go back to Arthur Eddington and Bertrand Russell and can be found in many contemporary panpsychist authors. See William E. Seager, "The 'Intrinsic Nature' Argument for Panpsychism," *Journal of Consciousness Studies* 13, nos. 10–11 (2006): 129–145.

89. See Westerhoff, *Nāgārjuna's Madhyamaka*, chap. 2.

90. For Madhyamaka-inspired arguments in analytical philosophy that there cannot be intrinsic properties, see Westerhoff, *The Non-Existence of the Real World*, 195–213. For quantum mechanics, see Carlo Rovelli, *Helgoland: Making Sense of the Quantum Revolution* (New York: Riverhead Books, 2021), 148–158.

91. Keith Frankish, "The Mental Life of Mountains," *New Humanist*, April 22, 2022, https://newhumanist.org.uk/articles/5951/the-mental-life-of-mountains.

92. The locus classicus of illusionism is Daniel C. Dennett, *Consciousness Explained* (Boston: Little Brown, 1991). See also Keith Frankish, "Illusionism as a Theory of Consciousness," *Journal of Consciousness Studies* 23. nos. 11–12 (2016): 11–39.

93. See Strawson, "The Consciousness Deniers."

94. Merleau-Ponty, *Phenomenology of Perception*, 52.

95. Merleau-Ponty, 5.

96. There is another kind of illusionism, however, which we can call Buddhist illusionism. It maintains, based on certain lineages within Buddhist philosophy, that both the subject-object structure of ordinary consciousness and the impression that consciousness has an intrinsic nature are experiential and cognitive illusions. Buddhist illusionism is not premised on physicalism and would be partly sympathetic

to our critique of the Blind Spot. Considering this kind of illusionism would take us beyond the scope of this chapter. See Jay Garfield, "Illusionism and Givenness," *Journal of Consciousness Studies* 21 (2016): 73–82. For a critical response, see Evan Thompson, "Sellarsian Buddhism: Comments on Jay Garfield, *Engaging Buddhism: Why It Matters to Philosophy*," *Sophia* 57 (2018): 565–579.

97. Chalmers, *The Conscious Mind*.

98. See Piet Hut and Roger Shepard, "Turning the 'Hard Problem' Upside Down and Sideways," *Journal of Consciousness Studies* 3, no. 4 (1996): 313–329; Francisco J. Varela, "Neurophenomenology: A Methodological Remedy for the Hard Problem," *Journal of Consciousness Studies* 3, no. 4 (1996): 330–349; Michel Bitbol, "Science as If Situation Mattered," *Phenomenology and the Cognitive Sciences* 1 (2002): 181–224.

99. Edmund Husserl, *On the Phenomenology of the Consciousness of Inner Time (1893–1917)*, trans. John Barnett Brough (Dordrecht: Springer, 1991); Alfred North Whitehead, *The Concept of Nature* (Cambridge: Cambridge University Press, 1920; Ann Arbor: University of Michigan Press, 1957).

100. Kathleen A. Garrison, Dustin Scheinost, Patrick D. Worhunksy, Hani M. Elwafi, Thomas A. Thornhill IV, Evan Thompson, Clifford Saron, Gaëlle Desbordes, Hedy Kober, Michael Hampson, Jeremy R. Gray, R. Todd Constable, Xenophan Papademtris, and Judson A. Brewer, "Real-Time fMRI Links Subjective Experience with Brain Activity during Focused Attention," *Neuroimage* 81 (2013): 110–118.

101. Hut and Shepard, "Turning the Hard Problem Upside Down and Sideways," 320–321.

102. Dunne et al., "Mindful Meta-Awareness: Sustained and Non-Propositional"; Antoine Lutz and Evan Thompson, "Neurophenomenology: Integrating Subjective Experience and Brain Dynamics in the Neuroscience of Consciousness," *Journal of Consciousness Studies* 10 (2003): 31–52; Sina Fazelpour and Evan Thompson, "The Kantian Brain: Brain Dynamics from a Neurophenomenological Perspective," *Current Opinion in Neurobiology* 31 (2015): 223–229.

103. See Claire Petitmengin and Jean-Philippe Lachaux, "Microcognitive Science: Bridging Experiential and Neuronal Microdynamics," *Frontiers in Human Neuroscience* 27 (2013), https://doi.org/10.3389/fnhum.2013.00617.

104. See Christopher Timmerman, Prisca R. Bauer, Olivia Grosseries, Audrey Vanhaudenhuyse, Franz Vollenweider, Steven Laureys, Tania Singer, Mind and Life Europe (MLE) ENCECON Research Group, Elena Antonova, and Antoine Lutz, "A Neurophenomenological Approach to Non-Ordinary States of Consciousness: Hypnosis, Meditation, and Psychedelics," *Trends in Cognitive Sciences* 27, no. 2 (2023): 139–159, https://doi.org/10.1016/j.tics.2022.11.006.

105. See Thompson, *Waking, Dreaming, Being*.

Chapter 9

1. See Carolyn Merchant, *The Death of Nature: Women, Ecology, and the Scientific Revolution* (New York: HarperOne, 1980).

2. Vladimir Vernadsky, as quoted in A. V. Lapo, "Problemy biogeokhimii" ["Problems of Biogeochemistry"], *Works of the Biogeochemical Laboratory* 16 (1980): 123, http://scihi.org/vladimir-vernadsky-biosphere/.

3. Vladimir I. Vernadsky, *The Biosphere*, trans. David B. Langmuir (New York: Springer Science + Business Media, 1998), 44.

4. Vernadsky, 56.

5. Quoted in Alexej M. Ghilarov, "Vernadsky's Biosphere Concept: An Historical Perspective," *Quarterly Review of Biology* 70 (1995): 197.

6. Ghilarov, 196.

7. James Lovelock, *Homage to Gaia: The Life of an Independent Scientist* (New York: Oxford University Press, 2000), 255.

8. See James Lovelock, *Gaia: A New Look at Life on Earth* (New York: Oxford University Press, 1979), 49. For discussion of the role cybernetics played in the Gaia theory, see Bruce Clarke, *Gaian Systems: Lynn Margulis, Neocybernetics, and the End of the Anthropocene* (Minneapolis: University of Minnesota Press, 2020).

9. James E. Lovelock, "Gaia as Seen through the Atmosphere," *Atmospheric Environment* 6 (1972): 579–580. See also Lovelock, *Gaia*.

10. See Lynn Margulis, *Symbiosis in Cell Evolution: Microbial Communities in the Archean and Proterozoic Eras*, 2nd ed. (New York: Freeman, 1992); Lynn Margulis, *Symbiotic Planet: A New Look at Evolution* (New York: Basic Books, 1998).

11. Robert J. Charlson, James E. Lovelock, Meinrat O. Andreae, and Stephen G. Warren, "Oceanic Phytoplankton, Atmospheric Sulphur, Cloud Albedo and Climate," *Nature* 326 (1987): 655–661.

12. See Lynn Margulis and Dorion Sagan, *What Is Life?* (New York: Simon and Schuster, 1995). See also Clarke, *Gaian Systems*.

13. For further discussion, see Evan Thompson, *Mind in Life: Biology, Phenomenology, and the Sciences of Mind* (Cambridge, MA: Harvard University Press, 2007), 119–122.

14. James W. Kirchner, "The Gaia Hypothesis: Fact, Theory, and Wishful Thinking," *Climatic Change* 52 (2002): 391–408.

15. Timothy M. Lenton, Stuart J. Daines, James G. Dyke, Arwen E. Nicholson, David M. Wilkinson, and Hywel T. P. Williams, "Selection for Gaia across Multiple Scales," *Trends in Ecology and Evolution* 33 (2018): 633–645.

16. Will Steffen, Katherine Richardson, Johan Rockström, Hans Joachim Schelln-huber, Opha Pauline Dube, Sébastien Dutreuil, Timothy M. Lenton, and Jan Lub-chenco, "The Emergence and Evolution of Earth System Science," *Nature Reviews Earth and Environment* 1 (2020): 54.

17. Steffen et al., 54.

18. Svante Arrhenius, *Worlds in the Making: The Evolution of the Universe*, trans. H. Born (New York: Harper, 1908).

19. Roger Revelle and Hans E. Suess, "Carbon Dioxide Exchange between Atmo-sphere and Ocean and the Question of an Increase of Atmospheric CO_2 during the Past Decades," *Tellus* 9 (1957): 18–27.

20. Quoted in Dale Jamieson, *Reason in a Dark Time: Why the Struggle against Climate Change Failed—and What It Means for Our Future* (New York: Oxford University Press, 2014), 20.

21. Paul J. Crutzen and Eugene F. Stoermer, "The Anthropocene," *Global Change Newsletter* 41 (2000): 17–18.

22. Peter M. Vitousek, Harold A. Mooney, Jane Lubchenco, and Jerry M. Melillo, "Human Domination of Earth's Ecosystems," *Science* 277 (1997): 494–499; Steven W. Running, "A Measurable Planetary Boundary for the Biosphere," *Science* 377 (2012): 1458–1459. For discussion, see Jamieson, *Reason in a Dark Time*, 178–179.

23. Vitousek et al., "Human Domination of Earth's Ecosystems," 494.

24. Clarke, *Gaian Systems*, 256.

25. Kathleen Dean Moore, *Great Tide Rising: Towards Clarity and Moral Courage in a Time of Planetary Change* (Berkeley, CA: Counterpoint, 2016), 132.

26. See Andreas Malm and Alf Hornborg, "The Geology of Mankind? A Critique of the Anthropocene Narrative," *Anthropocene Review* 1 (2014): 62–69.

27. Kyle Whyte, "Indigenous Climate Change Studies: Indigenizing Futures, Decol-onizing the Anthropocene," *English Language Notes* 55 (2017): 159.

28. Jamieson, *Reason in a Dark Time*.

29. Lukas Rieppel, Eugenia Lean, and William Deringer, "Introduction: The Entan-gled Histories of Science and Capitalism," *Osiris* 33 (2018): 4.

30. Rieppel et al., 2.

31. Rieppel et al., 2. For feminist scholarship, see Donna Haraway, "Situated Knowl-edges: The Science Question in Feminism and the Privilege of Partial Perspective," *Feminist Studies* 14 (1988): 575–599; Sandra G. Harding, *Whose Science? Whose Knowledge? Thinking from Women's Lives* (Ithaca, NY: Cornell University Press, 1991);

Evelyn Fox Keller, *Reflections on Gender and Science*, 10th ann. ed. (New Haven: Yale University Press, 1995).

32. Rieppel et al., "Introduction," 5.

33. Harold J. Cook, "Sciences and Economies in the Scientific Revolution: Concepts, Materials, and Commensurable Fragments," *Osiris* 33 (2018): 43.

34. Cook, 43.

35. See Amitav Ghosh, *The Great Derangement: Climate Change and the Unthinkable* (Chicago: University of Chicago Press, 2016). See also Whyte, "Indigenous Climate Change Studies."

36. William Deringer, *Calculated Values: Finance, Politics, and the Quantitative Age* (Cambridge, MA: Harvard University Press, 2018), xi.

37. Deringer, 6.

38. Deringer, xi.

39. Deringer, xi.

40. Deringer, 6.

41. William Nordhaus, "Projections and Uncertainties about Climate Change in an Era of Minimal Climate Policies," *American Economic Journal: Economic Policy* 10 (2018): 333–360.

42. Jamieson, *Reason in a Dark Time*, 6, 142.

43. Jamieson, 6, 143.

44. Jamieson, 143.

45. Jamieson, 143.

46. Stefan Thurner, Rudolf Hanel, and Peter Klimek, *Introduction to the Theory of Complex Systems* (New York: Oxford University Press, 2018), 1.

47. See Stuart A. Kauffman, *Investigations* (New York: Oxford University Press, 2000), and Stuart A. Kauffman, *A World Beyond Physics: The Emergence and Evolution of Life* (New York: Oxford University Press, 2019).

48. Kauffman, *Investigations*.

49. Thurner et al., *Introduction to the Theory of Complex Systems*, 15.

50. Alfred North Whitehead, *Process and Reality* (New York: Free Press, 1978), 289.

51. David K. Campbell, "Fresh Breather," *Nature* 432 (2004): 455.

Index